Wild China

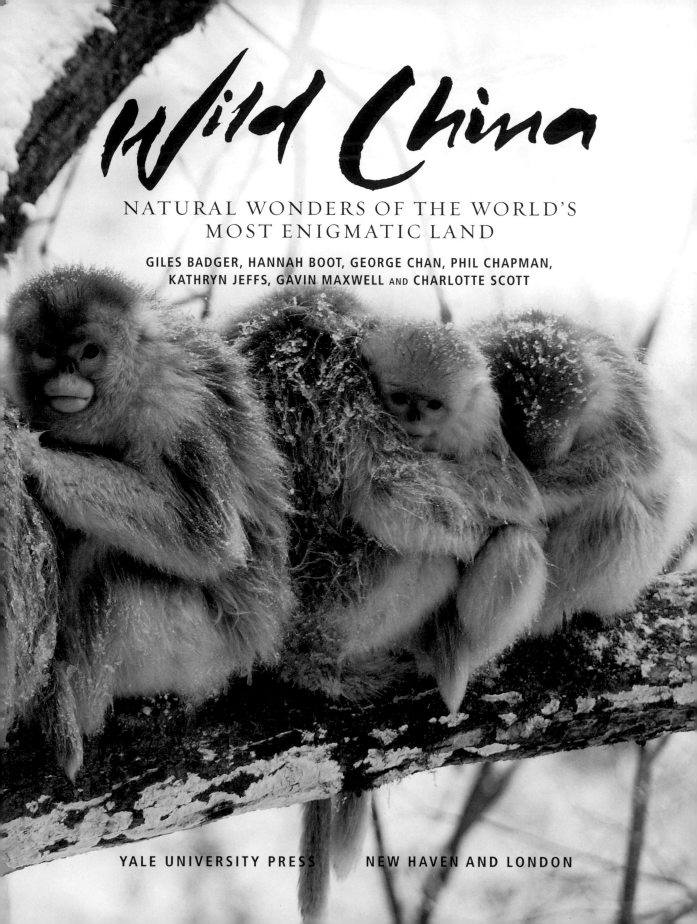

Wild China

NATURAL WONDERS OF THE WORLD'S MOST ENIGMATIC LAND

GILES BADGER, HANNAH BOOT, GEORGE CHAN, PHIL CHAPMAN,
KATHRYN JEFFS, GAVIN MAXWELL AND CHARLOTTE SCOTT

YALE UNIVERSITY PRESS   NEW HAVEN AND LONDON

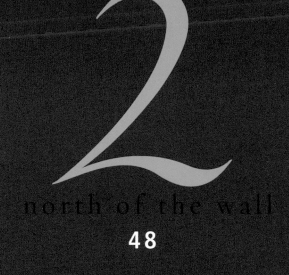

introduction

'WE HAVE TO ACCEPT THE FACT THAT THE BAIJI IS EXTINCT,' announced August Pfluger, joint leader of an expedition that, in 2006, surveyed the Yangtze River in a fruitless search for the rare freshwater dolphin. 'It is a tragedy, a loss not only for China, but for the entire world.'

The baiji, or Yangtze River dolphin, has long been an emblem of endangered wildlife in China. It was thought by scientists to have evolved as a separate species 20 million years ago in the turbid waters of the Yangtze, its only home. In the murky water, the long-snouted, pale-skinned dolphin depended almost entirely on its highly developed acoustic sense to navigate and catch fish. In recent decades, the increasing noise of ever greater numbers of ever larger motor vessels on China's premier inland waterway must have made its life unbearable. As if that wasn't bad enough, fishermen trapped baiji in their nets, and cargo ships and tourist boats sucked them into their propellers. Overfishing and industrial pollution reduced the baiji's food supply and poisoned its river home, and dams blocked its passage and altered its environment. The baiji didn't stand a chance.

The baiji may be the first large Chinese mammal species in historic times to have been lost for ever as the result of human activity. But will it be the last?

China is a land of more than 1.3 billion people, with the world's most rapidly growing economy, largely based on manufacturing and agriculture. Travelling in China while making the BBC *Wild China* series, we were constantly amazed at the sheer scale and pace of urban development and of construction – highways, high-rise cities, civil-engineering projects, factories and power stations. The first impression is of a society bent on economic advancement whatever the price and a country where nature is in full retreat. But there is another side to China.

ABOVE
One of the last pictures of one of the last baiji – the Yangtze River dolphin.

PREVIOUS PAGES
The great Hengduan ranges of Yunnan and Sichuan. The high, snow-covered mountain tops and the deep river gorges have separated and isolated populations of plants and animals, resulting in a huge diversity of species.

the big picture

With a land area almost that of the US, China still has significant areas where human density is relatively low. These represent a huge range of habitats, from the world's highest mountains to its most northerly deserts, from ice-bound winter seas to tropical coral reefs, from steaming subtropical rainforest to arid rocky basins, together with fabulous temperate woodlands, bamboo groves, grassy steppes, evergreen taiga forests, tundra, mysterious caves and smoking volcanoes.

The Yangtze is the third longest river in the world, at 6300km (3914 miles); and the Yellow River – the second longest in China – flows for about 5500km (3420 miles). Of the world's 19 mountains higher than 7000 metres (22,965 feet), 7 are in China. The Qinghai-Tibet plateau, the world's highest, borders onto the world's highest mountain peak, Everest (Qomolangma). The canyon of the Yarlung Tsangpo River, the 5382-metre-deep (17,657-foot) Yarlung Gorge, is arguably the world's largest and deepest – three times deeper than the Grand Canyon in Arizona.

Tucked away within this incredibly diverse landscape is a wealth of animal and plant life. China is home to 534 species of mammals – one eighth of the world total – of which more than a hundred, including iconic species such as the giant panda, evolved here and live only here. China's birdlife is also extremely rich, with more than 1300 species, and there are more than 2200 species of fishes.

There are an amazing 32,800 species of higher plants – an eighth of the world total – making China the third richest country in the world for plantlife after Malaysia and Brazil. Many of the most beautiful cultivated plants in temperate gardens worldwide are native to China, including azaleas, camellias, rhododendrons, forsythias, clematis, dogwoods and polyanthus. China is also a centre of diversity for food and medicinal plants including peaches, oranges, lemons, grapefruit, walnuts, chestnuts, apples, lychees, rice, barley, soya beans, tea, ginger and liquorice.

ABOVE
A wild hydrangea from Yunnan. Many of our familiar garden plants originate from China.

As a predominantly rural society, the art, literature and traditions of China have long celebrated the beauty of its landscapes, birds and flowers. The Chinese are proud of their natural riches, and many of China's notable plants and animals are officially protected in national parks and wildlife reserves.

OPPOSITE
Black-necked cranes
wintering on
farmland near Lhasa,
Tibet. Record
numbers are now
wintering on the
Tibetan Plateau.
China has eight
species of cranes –
symbols of long life
and happiness.

But official protection is not a guarantee of safety. Historically, conservation has been poorly funded and protection has been weak. Many reserves, especially locally designated ones, have been understaffed. Reserve staff in the main have had little specialist knowledge or expertise and have been ill-equipped to deal with well-organized groups of outsiders who come to hunt, mine or collect plants in nominally protected areas. Illicit hunting is a particular problem, given China's long tradition of exploiting its wild animals and plants for food and medicine.

attitudes to nature

Traditional Chinese medicine is a holistic approach to healthcare, involving treatment of patients using plant, mineral and animal-based ingredients, including tiger bone, rhino horn and bear bile. Though many species used are protected by national and international laws, illegal trade and poaching have increased to crisis levels as the population has got more urban and more affluent. Guangxi province has a state-sponsored tiger-breeding centre that keeps upwards of a thousand tigers used to produce tiger-bone wine.

'We will eat anything with legs, except a table, and anything with wings, except a plane,' goes the famous southern-Chinese adage. In many areas, the array of cages outside the local restaurant may display a wider range of wildlife than a visitor can hope to see in the wild. Such wild food is valued above farmed produce – regarded as promoting health, being especially good for virility and a strong immune system. At formal banquets, which are a key part of many business transactions, unusual and therefore rare and expensive animals are particularly prized as centrepieces to the feast, as a way to honour important guests.

Current legislation allows the consumption of about 50 species of wild animal if they are farm-raised. But the distinction is easily blurred. Despite attempts to regulate and control the open sale of wild food in China, it remains big business. There are no official statistics on the amount of wildlife being eaten in China, but the China Wildlife Conservation Association estimates that in Guangdong province alone, 50 tons of wild frogs, 1000 tons of snakes and several thousand tons of wild birds are consumed in stores and restaurants each year, plus an array of mammals including badgers, civets, bats and pangolins.

Not all cultural practices in China are destructive to wildlife, and one of the highlights of filming the *Wild China* series was discovering cultures that valued wild animals in a different way. These included the nomadic reindeer-herders of northern Heilongjiang, the Miao villagers with their weather-forecasting swallows in Guizhou, the Kazakh eagle-hunters in Xinjiang, and the pheasant-feeding Tibetan monks. And looking at the history of China, such respect for other life forms is not surprising, given that Daoism and Buddhism are so much a part of Chinese culture.

While filming stories about the relationship between people and wildlife in China proved relatively straightforward, our *Wild China* team found it much more difficult to capture sequences of wild animal behaviour. Studies of animal behaviour are not common in China, and very few groups of wild animals are habituated to the presence of human observers enough to be approachable by naturalists or professional film-makers. Those that are approachable are so strictly protected that it is not easy for outsiders to gain permissions to see them, let alone to film them. Less heavily protected animals tend to be wary of people and so are very hard to observe or film. Many species records confidently listed in the faunal inventories of nature reserves prove to be old sightings, and the animals either no longer occur or are vanishingly rare. There is also little accessible information for foreigners about where or how to see wild creatures, though doubtless local people and hunters will be knowledgeable.

But when compared to countries such as the UK, which has already killed off most of its large animals, China retains an impressive list of wildlife. The world's most populous country still has iconic large mammals such as elephant, gaur, takin, giant panda, leopard, snow leopard, black and brown bears, wild yak, Bactrian camel and even a few Amur (Siberian) tigers.

ABOVE
Limestone formations surrounded by huge plantations of oilseed rape in Qujing, southern Yunnan. Much farming in China is now on an industrial scale.

the future

China was one of the first countries to ratify the Convention on Biological Diversity, in 1993. Since then, it has catalogued national species, created an endangered list, shown determination to enforce the Convention on International Trade in Endangered Species (CITES), and designated many species and habitats as 'protected'. There are now 2194 nature reserves in China covering some 15 per cent of the country – above the world average of 11.5 per cent and an area six times that of Great Britain.

Modern China is emerging from a tumultuous period in its history with a sense of purpose, pride and optimism. We found a rapidly growing environmental

consciousness among Chinese young people, expressed as growing support for budding grass-roots conservation organizations. The media in China also increasingly carries environmental or conservation-related stories, raising interest and awareness.

Today's China remains a land of extraordinary beauty, with breathtaking landscapes and a wonderful array of wildlife. Given the determination to protect its wildlife through the current phase of economic expansion, there is a chance for China's natural wonders to remain a source of pride and wonder for many generations to come.

BELOW
The Great Wall – the world's longest man-made structure and a UNESCO World Heritage Site. Just as impressive are China's natural wonders, though many need greater protection.

the
heartland

IN 1793, A BRITISH ENVOY TRAVELLED TO THE HEART OF CHINA to explore potential trade agreements. Upon receiving an audience with the emperor, he was curtly told: 'We possess all things and have no use for your country's manufactures.' The Chinese believed they had little need for the outside world. This largely self-reliant outlook meant that China's heartland remained comparatively off-limits to foreigners until relatively recently. Images of the Forbidden City and tales of pandas and dragons all contributed to the intrigue. Protected and shielded by the Great Wall in the north, forbidding mountains of the Tibetan plateau in the west, the sea in the east and tropical jungles in the south, China's heartland has been culturally and geographically separate for millennia.

China has the longest unbroken recorded history in the world. More than 4500 years ago, Huang Di, the legendary Yellow Emperor, united neighbouring tribes in the Yellow River (Huang He) basin on the north China plain and today is regarded as the father of the Chinese civilization. But it was not until 221 BC that the government was centralized and the empire of China created by Emperor Qin (of terracotta-army fame). The western spelling of Qin – 'Ch'in' – gave rise to the name China. This period was followed by the Han dynasty, a golden age of cultural flowering, when writing, art, mathematics, science, music and medicine flourished. It crystallized the Chinese people's identity as a race, and the Han name persists to this day.

The Yellow River heartland quickly expanded as the fast-growing and hungry civilization spread southwards towards the Yangtze River (Chang Jiang). The fertile plains of these rivers became China's 'breadbaskets', with wheat in the temperate north and rice in the subtropical south – the Qinling Mountains dividing the two like a giant backbone. Today, the Han people comprise 92 per cent of China's one-billion-plus population – the largest ethnic group in the world. And their language, Mandarin, is the world's oldest living language.

the mother river

In 2200 BC, Emperor Yu the Great pronounced: 'Whoever controls the Yellow River controls China.' But for most of China's long history, the Yellow River has been in control. This is reflected in the names for the river: described as the 'Mother River of China' but also as 'China's sorrow'. The river has been and still is as vital to China as the Nile is to Egypt. But the relationship is a complex and difficult one.

Originating in the west of China on the high Tibetan plateau, the Yellow River flows eastwards through nine provinces before emptying in the Bo Hai Gulf, making it one of the longest rivers in the world at about 5500km (3420 miles). Along this route, it carves through a vast loess plateau, and the huge amount of loose yellow sediment that it carries away is what gives the river its name. So fertile is its cargo of sediment that it fuelled the growth of the Chinese civilization around the river's flat

basin. The fine silt steadily builds up the riverbed until the river bursts its banks and floods in dramatic and devastating fashion. Not surprisingly, China has always had someone responsible for flood control.

Projects on a massive scale were undertaken to irrigate the land and contain the river. In 246 BC, using only the simplest of tools, a canal project supplied 80,000 hectares (197,680 acres) around Xi'an, home to the Emperor Qin. Using dykes, canals, levees and artificial flood basins, the Chinese spent thousands of years continually fine-tuning the wayward river. The river's flow is so unpredictable, though, that containment has never been a total solution, and Chinese history is littered with accounts of drastic floods. 179 BC was known as 'the year of the great flood', and even in recent times, the statistics are shocking: in 1887, up to 2 million people died as a result (direct and indirect) of flooding, followed by nearly 4 million in 1931. Each time the river floods over the level plain, it reinvents itself, forging a new route. In the past 2000 years, the river has had 26 major changes in course.

m y s t i c a l   m o u n t a i n s

The fertile river basins have been populated by people for thousands of years, steadily pushing the wildlife up to the mountains that cut and thrust their way through the heartland. These mountains remain sanctuaries for wildlife, not only because of their relative inaccessibility but also because they have long been viewed as sacred. Early beliefs suggested that mountains were pillars that held heaven aloft and prevented it flattening the Earth. Mountains became places of pilgrimage and worship, reputed to be inhabited by shamans and mystics, who survived on a diet of magic herbs that enabled them to live for hundreds of years. Since 219 BC, Tai Shan mountain in the north has been visited by Chinese emperors, who believed it to be a god – the son of the Emperor of Heaven. At the other end of the heartland, Emei Shan in Sichuan province was the site of the first Buddhist temple in China, founded 2000 years ago.

The monks at Mount Emei are not the only inhabitants: the misty peaks are home to impressive macaques weighing up to 25kg (55 pounds) – the biggest in the world. They forage mostly on fruit, but the monks also feed them. The combined spectacles of landscape, religion and wildlife draw thousands of pilgrims and tourists, which means more easy food for the macaques. In fact so aggressive are some of these monkeys in their demands for food that there are tales of people being chased off the edges of the cliffs.

Despite this, many believe a good deed towards the human-looking primates is an investment in eternity.

Another big tourist destination lies on the edge of the Tibetan Plateau at the western border of the heartland. Jiuzhaigou is a remote region of rugged limestone mountains and dramatic forested valleys, home to Tibetans. It wasn't explored by outsiders until the 1970s and has since been declared a World Heritage Site. Water cascades down the mountains forming beautiful lakes. More than a hundred of these are filled with calcium-rich water of unbelievable clarity and extraordinary colours. Ghostly white underwater forests of calcium deposits are home to a type of carp found only here.

In autumn, the leaves are a blaze of colour, contributing to a view of Jiuzhaigou as a 'paradise on Earth'. But the real treasure trove of China's sacred and magical mountains has to be the Qinling range, which divides the temperate north and the subtropical south. The result of straddling the two regions is a wealth of diversity, with leopards, golden monkeys and giant pandas. But as the human population in the heartland has grown, so the Qinling Mountains have become increasingly isolated.

OPPOSITE
Mirror Lake, one of many beautiful lakes in the Jiuzhaigou (Nine Village Valley) World Biosphere Reserve and World Natural Heritage site in Sichuan, discovered by 'outsiders' only in the 1970s. The forests are a mix of deciduous and coniferous trees, with an understorey of bamboos and rhododendrons. Takins and giant pandas are still found here in small numbers.

the legendary white bear

In 1869 missionary and naturalist Père Armand David first set eyes on a legendary animal known as the 'white bear'. He subsequently wrote: 'I have not seen this species, which is easily the prettiest kind of animal I know, in the museums of Europe. Is it possible that it is new to science?' Today, this 'new' animal is the most instantly recognizable creature on the planet. It has, though, taken scientists more than a hundred years to figure out whether it is a bear or a raccoon.

Bears, raccoons and dogs evolved from a common carnivorous ancestor that prowled around more than 24 million years ago. But though placed in the carnivore family, the panda is a herbivore. At 1.5 metres (4.9 feet) long and weighing up to 135kg (298 pounds), it may look like a bear, but its skull and teeth are different from those of bears. It has no heel pad on its hind feet, it doesn't hibernate, and even its genitalia are not bear-like. But the latest DNA studies reveal the panda's closest relative to be the spectacled bear of South America – so a bear it is after all.

The function of its distinctive coloration has been another subject of debate. Most of the time, pandas lead solitary lives, and so perhaps the colouring helps them spot each other in the dense forests and avoid unnecessary encounters. Or it may provide camouflage in the snow. The black eyes and ears are believed to make it look intimidating – black ears perhaps imply two sets of eyes. A dominant panda in an encounter may bob its head, emphasizing the impact of its eyes and ears, while a submissive one will turn its head to one side and cover its eyes with its paws. What is certain is that the giant panda is a survivor, having lived for at least 3 million years in the isolated bamboo forests of China.

PANDA LOVE

the female seeks out a remote and safe cave in which to give birth to a tiny, toothless, blind and naked baby

In spring, for a brief time, the strange bleating calls of amorous pandas can be heard in the forest. The peak receptive time for a female lasts just two to seven days, and competition for her favours is fierce. Several scent-marking males may pursue one female, fighting and even drawing blood, though size normally wins out. After mating, the female seeks out a remote and safe cave in which to give birth to a tiny, toothless, blind and naked baby. The baby is so tiny – just over one thousandth of its mother's weight – it holds the record for the smallest infant-to-mother ratio in placental mammals (those other than marsupials and monotremes). It will remain with her for 18 months.

Bamboo both defines the panda's nature and is the primary cause of its vulnerability. Bamboo is tough, fibrous and low in energy. Yet despite the panda's carnivorous gut, its diet is almost 99 per cent bamboo. Perhaps competition from other carnivores forced it to take up the bamboo lifestyle and evolve accordingly. Flexible forepaws and an enlarged wrist bone work like a primate's opposing thumb to help handle bamboo. Extra large molars and jaw muscles help crush it – muscles that give rise to the panda's big, round face – and a tough lining protects the throat and stomach. But the short carnivore intestine means it can digest only about 20 per cent of what it eats (a cow can digest about 60 per cent).

Panda faeces are so full of undigested material that, when fresh, they are an unmistakable bright green, and a panda has to spend up to 16 hours a day eating. Unlike bears, it cannot store enough fat to see it through winter hibernation, but its coat is impregnated with waterproof oil and has two layers – a coarse outer one and a dense woolly inner one – to help it endure freezing temperatures.

For much of the 1970s and 1980s, intensive logging drastically reduced the panda's chances of survival, and today its range is curtailed still further by farmland at the foot of the mountains. As numbers decreased, attempts were made to try to breed pandas in captivity – with limited success due to a lack of understanding of

panda behaviour. But the situation is changing, and a logging ban and improvements to the forest environment have increased its chances of breeding in the wild.

Research in the wild has informed captive-breeding programmes, and in the past ten years, there has been an extraordinary turnaround for both wild pandas and the captive-bred pandas at Wolong in Sichuan. Captive pandas are now given physical training to make them fit enough to perform, and they have even been subjected to videos of pandas reproducing – useful for the sound effects. In 2006, Wolong reared 17 cubs, some of which may one day be released into the wild. More important, the wild population has grown to about 1600 animals.

the golden creatures

In the sixteenth century, in Constantinople, there lived a beautiful slave girl named Roxellana. Her striking red hair, large blue eyes and exquisite upturned nose won her the attention of the Sultan of the Ottoman Empire, who was so besotted with her that he married her. Today, she is immortalized in the scientific name *Rhinopithecus roxellana*, given to a striking animal living in the same forests as the

ABOVE
A rare shot of a wild giant panda. Only 29 small, fragmented areas remain that have enough habitat to support populations of pandas.

OPPOSITE
A male giant panda attempts a handstand as he scent-marks a trunk. It's a message for rival males – the higher the mark, the bigger the panda.

panda. Measuring up to 2 metres (6.6 feet) from head to tail, it has long, thick, golden red hair, a dazzling sky-blue face set off with fleshy cheek flaps (in the case of males) and a nose turned up to a peak. Living at altitudes of 1500–3500 metres (4921–11,483 feet) and with ranges of up to 50 square kilometres (19.3 square miles), golden snub-nosed monkeys were seldom encountered, and tufts of their golden fur were the source of much intrigue, regarded by some as evidence of the existence of the legendary yeti. But in the past decades, more has been learnt about their lifestyle.

In the high valleys of the Qinling Mountains, they can be heard leaping along the hillsides. A troop of up to 300 monkeys can be surprisingly noisy, with a chatter of bickering, grooming and copulating, occasionally erupting in howls and whoops as sub-adult males attempt to challenge the hierarchy. In summer, they eat leaves, the bacteria in their sacculated stomachs helping to break down the cellulose. In winter, as the snow begins to fall, the troops split into smaller groups and survive on bark, lichen and moss. After a bout of foraging, the monkeys will retire to a sunny spot to spend a lazy few hours digesting, each family unit snuggled together to reinforce social bonds as well as to stay warm. Thanks to their thick fur, they can survive in temperatures that would kill many other monkeys. In the past, Manchu officials wore their pelts, as their fur was believed to prevent rheumatism. Today, however, golden monkeys are a protected species, with a seven-year prison sentence for anyone caught hunting them.

Another charismatic creature shares the valleys of the Qinling Mountains with the golden monkeys. Picture an animal with a golden fleece, the horns of a wildebeest, a shaggy head with an arched Roman nose, the body of a musk ox and a tiny tail. In an attempt to make sense of it, the golden takin is described as a 'goat antelope', but this doesn't convey the fact that it is one of the most formidable creatures in China. Should you encounter one, you are advised to climb the nearest tree.

The takin is from the same family as the Arctic musk ox. Males are 1.3 metres (more than 4 feet) at the shoulder, 2.2 metres (more than 7 feet) long, weigh 350kg (772 pounds) and are armed with 30cm (12-inch) horns. Its appearance is enhanced by a thick, shaggy golden coat that is rumoured to be the golden fleece of Jason and the Argonauts fame. Though the fleece is exotic enough to warrant the name, it also smells, being impregnated with an oil that helps insulate the takin against winter temperatures that can drop to -10°C (14°F).

Takins live in dense bamboo forests high up in the mountains where the terrain is rugged and treacherous. But their broad hooves and large dewclaws make them surprisingly nimble, and they migrate up and down the Qinling mountainsides following their seasonal food. In winter they eat twigs and evergreen leaves, standing on hind legs or pushing over saplings, and in summer they graze in alpine meadows and browse on deciduous leaves.

BELOW
A herd of golden takins grazing on the lower mountain slopes in summer. Takins will attack if threatened and so are feared by people, which may have helped their survival.

BELOW
Golden snub-nosed monkeys in huddles,
reinforcing social bonds. In winter, they
survive by foraging on lichen, bark and moss
and will huddle together to keep warm.

thanks to their thick fur, they can survive in temperatures
that would kill many other monkeys

The big males are usually solitary, but in summer, herds of up to a hundred form in preparation for the rut. The males compete by charging each other and clashing horns, sending gunshot-like cracks reverberating around the mountains. The winner of these gladiatorial contests attracts a harem of females. Babies are born the following March, when they can fall prey to bears and wolves, which might explain the takin's aggressive nature. But the main threat is its dwindling habitat, and there is a real possibility that the endangered takin might, like the golden fleece, pass into legend.

return of the ibis

Not much more than 20 years ago, another strange-looking species came as close as can be to extinction. The crested ibis has a bright red face, whitish pink plumage that turns progressively pinker throughout the breeding season and an elongated curved bill for digging and stabbing snails and fish. The males have a crest that resembles native-American feathered headdress. Historically, crested ibis nested from Japan to the far east of Russia, but in 1981, there were just seven left in the wild – four adults and three chicks, living near Yangxian, a farming town on the southern slopes of the Qinling Mountains. Crested ibises need a combination of tall trees for nesting and roosting, and wetland – much of which has vanished, as wheat has replaced rice. The Yangxian paddy fields provided mini-wetlands, where the birds could hunt river snails and small fish, and a handful of trees that were crucial for nesting and roosting.

The crested ibis has always been associated with Chinese farmers – ancient poets observed how the noisy departure of the crested ibises from the roost heralded the start of the farmer's day. On discovery of the last remaining crested ibises, the Chinese embarked on a series of measures to protect them, including the banning of logging and the use of chemicals in the fields. The nest sites were declared state property and patrolled, and certain fields have been maintained through the winter to ensure a stable food supply. So successful has management and protection of the habitat been that today there are estimated to be about 360 individuals in the wild. Most of the world's population

LEFT
Roosting crested ibises. The surviving birds stay close to the paddy fields, which act as mini-wetland feeding grounds. Virtually the whole world population is found in this one area in the Qinling Mountains.

roosts on just a few trees, and every morning, as the farmers go to their fields on their bikes and tractors, the ibises fly off together. In the evening, as the farmers return home, the ibises return to their roost, flying in groups of dozens or more, a honking pink streak through the sky. The birds are still endangered, but they have become a symbol of hope and stability in this fast-changing country.

the water dragon

The fabulous creature that is quintessentially Chinese is the dragon – the Chinese have even called themselves Long de Chuanren, which means descendants of the dragon. Associated with the emperor as having fearsome and potentially destructive power, the dragon was replaced in modern times by the panda as China's national animal. Yet the dragon still exerts a powerful grip on the Chinese imagination.

It is not the fire-breathing monster that terrorized medieval Europe but a very different creature, which sheds light on the relationship of the Chinese to their world. The most obvious non-mythical relative is the extremely rare Chinese alligator, found along the Yangtze River in southern China. During the courtship season, the males roar and bellow, thrashing the water with head slaps and performing a spectacular water dance, which involves vibrating their bodies as they issue subsonic calls through the water to alert rivals and females of their presence, their tails arching and curving as they do so. Where the alligators live alongside rice fields, they are known as muddy dragons.

But the Chinese dragon never took a single definitive form – various depictions have included the head of a camel, horns of a deer, paws of a tiger and claws of an eagle.

Perhaps more surprising for those used to medieval dragons was the role it played not as malevolent monster but as a force for good: in the Chinese zodiac the appearance of the Draco constellation in spring was said to be when the dragon rose into the sky to herald the rainy season – the time of fruitfulness and growth. For a civilization dependent on the water of the Yellow River, the dragon was highly significant.

As a scaly creature, water was its natural element. The dragon blew steam, not fire, through its snout to create rain. Dragons that fought in the water caused floods, and storms were created by dragons fighting in the sky. Their long sinuous bodies carved out great rivers, and when coiled, these bodies became the dramatic shapes of mountains, which themselves are rainmakers.

Since dragons appeared to control the water, they were the source of immense fertility, knowledge and power. The dragon could ward off evil spirits and bring good fortune. Such was its role that Chinese emperors adopted its image for their robes and became associated with its power as they sat on their dragon thrones. You would ignore or insult the dragon at your peril. Temples had areas in which to worship and appease the dragon. Even today, defacing an image of a dragon is strictly taboo, and the dragon dance is still the climax of the 14-day Chinese New Year festival. Large cities put on spectacular displays of dragons, which parade through the streets accompanied by colourful lanterns, while in smaller villages, 'puppet' dragons run through the streets and into people's houses, where they are given gifts. The landscape, too, is still full of evidence of dragons, with endless dragon-named valleys, gorges, waterfalls and lakes. With modern China's ever-increasing dependence on water, and with the unruly Yellow River still on the agenda, it might be wise not to ignore the dragon.

schools of thought

China's unique perspective on its world has been forged by three main systems of thought. As early as the sixth century, a scholar named Li Shiqian observed, 'Buddhism is the sun, Daoism the moon, and Confucianism the five planets.' These three systems were less like religions and more like philosophies. The ultimate goal of each was for people to live in harmony with the world.

Confucius arrived first. His life (551–479 BC) coincided with a period of political turmoil, and his solution to his country's woes was a social structure in which all

LEFT
A gold-plated dragon creature from the Songzhanling Monastery. Dragons both ward off evil spirits and represent knowledge, fertility and power.

OPPOSITE
A student at the Confucian Temple of Literature at Jianshui, Yunnan. Formed in 1285, it is the second largest Confucian temple in China. Since the 1980s, more religious freedom has allowed many old beliefs and faiths to flourish.

people knew their place and what was expected of them. There were codes of conduct based on benevolence, etiquette and discipline. People were part of a series of relationships, from parents and siblings to spouses and business colleagues, all the way up to the emperor. Everything in life was relational, everyone was interconnected. One of his many catch phrases was 'do not do to others what you yourself would not desire.' While everyone was subject to fate, the path of Confucius expected everyone to strive for a better existence within his or her social rank instead of merely succumbing to a mapped-out life. This philosophy was, unsurprisingly, revered by successive emperors, and the teachings of Confucius were regarded as the 'state religion' until the changes wrought during the twentieth century.

At the same time, a very different school of thought was gaining acceptance. Daoism sought to escape from the routines and rituals of everyday life and strive for a simpler and more spiritual approach to existence. Its roots were based in the shamanistic and spirit worship of more ancient cultures. Nature was supreme, and everything was to be the way nature ordained. The social institution so championed by Confucius was irrelevant, and instead the Daoists imagined a utopian world with minimal laws and a return to a state of innocence. The concept of 'going with the flow', following the natural way and letting nature take its course was integral to Daoism. Nature itself existed in 'yin and yang' states – complementary opposites such as night and day, male and female, fire and water. Neither is absolute, but both are fluid and they must be held in balance. When viewed from other perspectives, everything can be broken down into yin and yang states, and balance can be achieved.

The writings and verse of Daoism are some of the most exquisite in Chinese culture, and this resulted in a dichotomy: for many educated Chinese, life had a social aspect where Confucian doctrine prevailed but also a private one with Daoist aspirations.

the arrival of buddhism

During the second century BC, the 'silk road' was becoming a well-travelled route to and from China, and it was how Buddhism, travelling via central Asia, made its way into China. Buddhism preached reincarnation, in contrast to the Chinese belief in a single life. It took time to win followers, and the fact that it was a foreign religion did not help. But with more periods of political turmoil, Buddhism seemed to offer explanations and comfort and began to capture people's imagination.

Buddhism promised a path to salvation through meditation and ascetic practices, with a strict code of morality. Its imagery and rituals were imbued with magic, and it quickly fused with Daoism – so much so that the founder of Daoism was said to have been reborn in India as the Buddha. The earliest Buddhist translations for the Chinese drew heavily on Daoist concepts and vocabulary to make the religion palatable. But the Buddhists still had to contend with Confucius.

Individual spiritual enlightenment and monasticism seemed at odds with social responsibility, the social order and a more practical way of living. But careful rewording reinvented Buddhism in a Confucian mould, whereby the salvation of an individual was for the greater good of society, and monks helped contribute to this greater good. Ancestor worship, a fundamental concept of Confucius, was incorporated, and Buddhist texts from India that emphasized family were given greater prominence in China. Once Buddhism was seen to be working for the

ABOVE
The eighth-century Buddha Maitreya (Buddha of the Future) on the Min River at Leshan, Sichuan. It's the world's largest carving of the deity.

Chinese system, it flourished. The essential unity of these three systems was stressed, rather than their differences exploited. It was said that Confucianism was consulted for family and ethical concerns, Daoism for physical and psychological concerns and Buddhism for matters relating to death and beyond.

Over the ensuing centuries, each had periods when they were in or out of favour, and all were eventually banned in the early twentieth century. Then in 1982, the constitution was amended to allow Chinese people considerable freedom of religious practice. Confucius is still ever-present, its legacy of system, society and obedience much in evidence. Daoism is the folk religion, and its many gods are still worshipped, especially in the countryside. Buddhism is the largest organized faith, involving up to 350 million people – 25 per cent of the Chinese population.

kung fu and the shaolin monks

Throughout history, few of China's philosophies and religions have aroused as much curiosity, wonder and notoriety as the Shaolin monks. In AD 540, a monk from southern India named Bodhidharma travelled to China to meet the emperor to discuss the finer details of Buddhism. But he fell out with the emperor. So he went to visit monks at a nearby temple, which had been built around a clearing planted with trees (Shaolin means young forest). Legend has it that the monks were too busy translating Buddhist texts to let him in, and so he retreated to a cave and began to meditate. Here he stayed for nine years before the monks finally realized he had something to offer – apparently his blue eyes had bored a hole in the wall of the cave. They let him in.

The monks had spent their lives sitting at desks, bent over in rapt concentration as they transcribed the texts, and so they were not in good shape and unable to perform all the physical and mental tasks required for Buddhist meditation. Many were unable to relax in the yogic postures, and some even fell asleep during meditation. The solution was to learn exercises in movement that would help restore energy, build strength and increase the flow of chi. Bodhidharma took as his inspiration iconic animals that would have been familiar to the Chinese. From these peculiar yet practical origins, Shaolin kung fu was born.

The exercises prescribed by Bodhidharma were mainly for energy and strength, and though the study of kung fu involved examining violence and its avoidance, it carried the caution that 'one who engages in combat has already lost the battle.' But the temple was wealthy and remote, and so it is possible marauding bandits were a problem and that this is why the monks perfected their practice into the martial art familiar the world over.

In AD 698, during an attempt to oust the emperor from his throne, the emperor's son was captured. Thirteen Shaolin monks were sent to rescue him.

Reputedly, they were victorious against an army of 10,000. This forged a bond between the emperor and the Shaolin monks that, for the most part, endured until the Ming dynasty (1368–1644), when they enjoyed their golden period. The monasteries were respected centres for philosophy, history, mathematics and poetry – essential ingredients required to make Shaolin kung fu masters.

Unfortunately, they fell out of favour in subsequent centuries, often finding themselves on the wrong side in a country racked by war and conquests. By the eighteenth century, they were outlawed and went underground. This is when tai chi flourished – its deliberately slowed movements were a kind of kung fu in disguise.

Shaolin re-emerged in 1895, following China's defeat by Japan and the threat of European invasion, when the Empress of China sanctioned the use of a previously banned society, the Righteous and Harmonious Fists, to cause havoc. More commonly known as the Boxers, this was a religious society built on Shaolin philosophy. They practised kung fu combined with magic rituals and spells that they believed made them impervious to bullets and pain. Unfortunately they were not, and the 'Boxer Rebellion' crumbled.

By the Cultural Revolution, the writing was on the wall for the Shaolin philosophy, which embraced Buddhism, Daoism and a system of teaching separate from Mao, and for kung fu – in China, at least. But thanks to Bruce Lee's films,

ABOVE
Warrior monks of the first-century Shaolin Temple, Dengfeng, displaying their kung fu skills at the temple's 'pagoda forest' (the tombs of eminent monks). The temple is the birthplace of kung fu, which is bigger in China today than it ever was, though the emphasis is more on the physical rather than philosophical elements that once inspired it.

THE KUNG FU MENAGERIE

the styles of
Shaolin kung fu
are based on
18 animals,
including cobra,
leopard, deer,
crane, mantis,
monkey and, of
course, dragon

The styles of Shaolin kung fu are based on 18 animals, including cobra, leopard, deer, crane, mantis, monkey and, of course, dragon. Each has it own particular style. The crane darts in and out, flapping its wings and stabbing with its beak, and the moves include long-range kicks and a hand formation called 'crane's beak' (*above*). Tiger is by contrast a furious medley of kicks, punches and grips – used as a last resort when cornered. Dragon is more subtle, relying less on physical projection and violence and more on a mental approach – unstoppable waves of energy. The one non-animal system of moves is 'drunk', a series of lurching steps and illogical blows that eventually confuse and exhaust the opponent.

kung fu has been huge everywhere else. It has only been recently that the concept of Shaolin kung fu has made a resurgence in a China eager to explore, and possibly exploit, its heritage. Perhaps wary of its chequered and challenging history, the emphasis is on the purely physical aspects and less on the spiritual and philosophical disciplines that once made it such a unique and devastating force.

the great wall

Self-defence, invasion and warfare have been ingredients of every civilization, but few have left such a mark on the landscape as has the Great Wall of China – the country's most iconic man-made structure. It marks the northern and western boundary of the heartland and is a rough divide between this largely temperate zone and more extreme landscapes and climates: the endless grasslands of Mongolia, the deserts of Xinjiang and the frozen tundra of the northeast that leads to Siberia.

The areas 'beyond the wall' were believed to be wild and dangerous, not only because of the land itself but also because of hostile people who lived there. From

700 BC, individual states in China had been building their own 'front lines' of walls to defend against different warring factions. From about 220 BC, when Emperor Qin united these states against common enemies in the north, the individual walls grew and connected and became one great wall. Much of it was constructed with earth, and so little of this structure remains today. It was not until a humiliating defeat by the Mongols in 1449 that the Great Wall began to take its present shape and course.

Rather than incorporating the distinctive bend of the Yellow River, the emperors of the Ming dynasty were forced to follow the southern edge of the Mongol-controlled Ordos Desert. It is one of the most audacious and extravagant feats of construction ever undertaken: 7.6 metres (25 feet) high and at up to 3.7 metres (12 feet) wide at the top, it was broad enough to act as a road for hundreds of marching troops and their supplies. Since much of the wall is in a state of disrepair and much is in remote and inhospitable locations, its true length is disputed – but estimates range from 2500km (1555 miles) to more than 6000km (3728 miles).

It may be vast, but contrary to popular belief, the Great Wall cannot be seen from the moon and would be barely visible to the naked eye even if you were orbiting the Earth. But its creation was testament to the extraordinary power of the emperor and the management of a civilization's resources. For as long as the Chinese empire was strong and united, the wall was generally effective in protecting its northern and western borders. But when China faced internal rebellion, the massive logistical effort of managing the wall suffered and its defence was quickly overcome.

BELOW
The Great Wall at Jingshanling. The treacherous terrain on either side of the wall would have been as much a deterrent to the invading hordes as the wall itself.

BELOW
The Great Wall, marking a boundary between the more temperate heartland and the great grasslands, deserts and frozen tundra of the wild northern and western regions.

it was not until a humiliating defeat by the Mongols
in 1449 that the Great Wall began to take its
present shape and course

During the Qing dynasty, Mongolia was finally annexed, and so the Great Wall was no longer needed as defensive barricade, and it fell into disuse. What it never stopped, though, were the waves of migrating owls, eagles, buzzards, cranes and other birds that fly south in a colourful and noisy procession each year to escape the Siberian-style winter of China's northeast.

invading sands, vanishing waters

Today the heartland faces a far more serious problem than human invasion: every spring, the north China plain is choked by dust storms that penetrate deep into the heartland. Deforestation in the north and west has given rise to persistent desertification, and the deserts are threatening to choke and swallow China's major cities – the nearest is now only 70km (43 miles) from Beijing. In an attempt to halt the invading sands, the Chinese have built a 'green wall' every bit as spectacular as the Great Wall and more vital than it ever was in the days of Genghis Khan. The 70-year project plans to plant a 4480km (2784-mile) belt of trees across the northwest rim of China in an attempt to stabilize the soil and halt the advance of the deserts.

Twentieth-century development has exacerbated many of the problems that have always confronted the heartland, not least those facing the mother river of China. Demand for water from the Yellow River has increased alarmingly: in 1950, the river irrigated 800,000 hectares (nearly 2 million acres) of farmland; in 2003, the area had increased to more than 7 million hectares (17 million acres). Such is the demand that, in recent years, the river has not been able to reach the sea.

A fast-growing population has meant that more people have moved onto the loess plateau, and the resulting deforestation and overcultivation have rapidly escalated the problems of loess erosion: 1.6 billion tons of silt were carried annually by the river in the 1950s; in the 1970s, this had risen to 2.2 billion tons.

There have been attempts to harness the river's power by constructing a series of huge dams. The success of these ventures has been distinctly variable – the Sanmenxia hydroelectric-power plant constructed in 1960 had no mechanism for dealing with the accumulated silt. Its turbines soon failed, and today the dam is no longer generating power; instead, it provides limited insurance against huge floods.

Meanwhile, nearly 2 million people have settled in the flood-diversion basins. Trying to control a river more than 5000km (3110 miles) long is a constant battle. Scientists believe barricades might hold for what are termed the '30 year floods', but historically, China also experiences one massive flood every century, and there is

BELOW
Construction of the Bird's Nest – the Olympic stadium in Beijing.

little provision to cope with this next onslaught. Upstream, though, reforestation and replanting have begun to stabilize the erosion of the loess plateau. When, in 2200 BC, the emperor pronounced 'whoever controls the Yellow River controls China', little did he know how relevant his words would be more than 4000 years later.

the temple of heaven

The modern metropolis of Beijing, with its high-rise office blocks, fast-food chains and bumper-to-bumper traffic, seems a world far removed from the ancient heartland. But it still has uniquely Han Chinese ingredients. Sprawling over a massive 2km (1.2 miles) in the heart of Beijing is the Temple of Heaven. Built in 1410, this is one of the most important sites in all of China. Since it represented heaven, the site was more than twice the size of the neighbouring Forbidden City, and at 275 hectares (680 acres), it's the largest centre of worship in the world.

The Temple of Heaven was the focal point for an entire civilization. Every winter solstice, emperors of the Ming and Qing dynasties, as the 'sons of Heaven', would lead an elaborate procession that travelled unseen by the masses from the Forbidden City to the temple. The emperor would come here to offer sacrifices and pray for a successful harvest – a risky proposition considering the fickle nature of the Yellow River but crucial to ensure the smooth running of the empire and to verify the emperor's role and total authority as the son of Heaven.

For hundreds of years, the temple was shrouded in secrecy and ritual, and it was not until the Chinese Republic in 1912 that the temple gates were finally opened to the multitudes. Today, thousands of Chinese visit the park, which plays a very different role from that of hundreds of years ago. Here they find a sanctuary. People perform tai chi, fly bird-shaped kites and practise singing – impromptu operatic renditions are a

BELOW
The Temple of Heaven, hidden from the masses for centuries but today one of China's most popular attractions. The temple grounds include a huge area of parkland.

TEMPLE OWLS

Emperors may no longer visit the temples, but the winter solstice is still marked by the arrival of equally secretive and charismatic creatures – the long-eared owls. They fly south every year from beyond the Great Wall. Some find sanctuary among the conifer trees that decorate the Temple of Heaven complex. They are hard to spot when roosting as they elongate their bodies and ears, compress their feathers and camouflage themselves as limbs of trees – a strategy probably designed to avoid predators. Groups will also gather close together – 11 in one tree is not uncommon.

For those with a practised eye, long-eared owls can be seen very close to the temple itself, facing south to make the most of the thin winter sun.

Standing 40cm (16 inches) tall, they are large, distinctive birds: their orange faces have a cross-shaped mark of white feathers, and their ear tufts point upwards as though constantly on the alert. Throughout most of the winter they remain silent, but when they do hoot, they can be heard up to a kilometre away. Like all owls they are magnificent silent hunters – their ears placed at separate heights to help detect their prey in total darkness before swooping down on silent wings. In Beijing, they probably dine out on rodents.

they fly south every year from beyond the Great Wall. Some find sanctuary among the conifer trees that decorate the Temple of Heaven complex

glorious distraction from the endless drone of Beijing traffic beyond the temple walls. Within many temple and park grounds, exquisite gardens are constructed with carefully placed rocks, pools and plants to create harmonious microcosms of nature.

The Han Chinese created a unique world hidden from outsiders for thousands of years. Within it they attempted to live a harmonious existence, founded on philosophies and arts that complemented not only the human world but also the natural and spiritual worlds. But with a massive and rapidly growing population, the demands on resources have reached a critical level. In theory, the Han Chinese have strived for harmony in all things, and they have left a rich legacy in the pursuit of this dream. In reality, this most noble of ambitions has been much harder to achieve.

BELOW
The interior of the wooden Hall of Prayer for Good Harvests at the Temple of Heaven. Four columns representing the four seasons support the domed ceiling without the aid of nails.

the emperor would come here to offer sacrifices
and pray for a successful harvest – a risky proposition
considering the fickle nature of the Yellow River

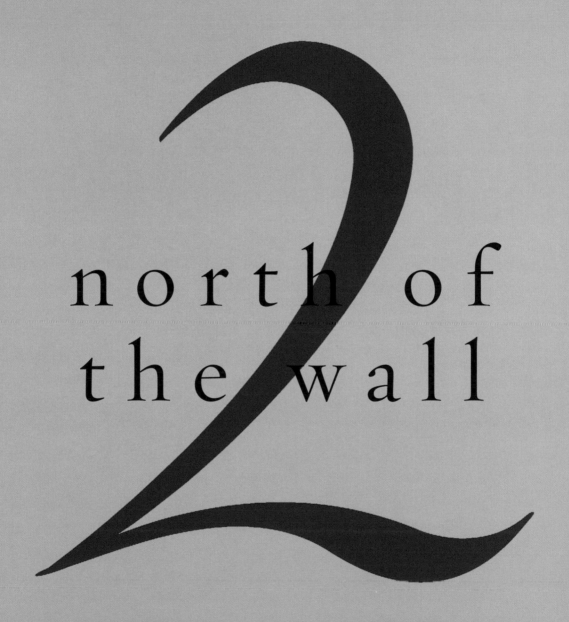

north of
the wall

2

AS YOU TRAVEL ACROSS NORTHERN CHINA, THE CLIMATE, geography, people, wildlife and landscape vary so much that the eastern and western corners hardly seem part of the same country. But China is no ordinary country. Nowhere else has a wall been built that divides the north from the south. North of the wall live the tribes that the Han people of China's heartland – the cradle of civilization – called the Hu, meaning the barbarians.

For thousands of years, these 'barbarians' – tribal nomads – have grazed their livestock over the plains and mountains, following the dramatically shifting seasons. Temperatures here can plummet to -52°C (-61°F) in winter when the Siberian winds blow. Summer temperatures can reach 50°C (122°F) or higher. Moisture from the East China Sea allows lush temperate forests to grow in the northeast, but as you travel west, it becomes increasingly dry, until you reach some of the most inhospitable deserts on Earth. All across northern China, people and wildlife have adapted to the harsh conditions that struck fear into the hearts of the Han people. Here survival depends on ingenuity and adaptability.

PREVIOUS PAGES
Domestic camels, one form of transport in the remote Badain Jaran, China's third-largest desert, lying within the Gobi.

OPPOSITE
Tian Shan, the Heavenly Mountains – a mighty wall of rock that forms the southern boundary of the Junggar Basin. Pastures on the northern slopes are watered by moisture blown in from Siberia.

freezing winters, monsoon summers

The highest and most famous mountain in northeast China, in Jilin province, lies on the border with North Korea. Its name, Changbai Shan, means ever-white mountain, and in winter it is covered in a dense blanket of snow. At its highest point, on Baiyun (white cloud) peak, you look down on Tian Chi, or Heaven Lake – the highest volcanic lake in the world at 2194 metres (7198 feet) above sea level. Under the snow, brown bears are hibernating, and most birds have flown south for winter. It is deathly quiet. But even in midwinter, when temperatures regularly drop to -40°C, the waterfalls and streams coming off the mountain are never quite frozen, heated by volcanic activity.

Changbai Shan lies near the western seaboard of China, and so for part of the year, it is bathed in moist monsoon air. The mountain slopes are rich in plantlife, with Mongolian oak and dwarf birch, leafless in winter, and Korean pine and Japanese yew, in leaf all year. Kiwi fruit grow here, and this area is famous for that most Chinese of medicinal plants – ginseng, now almost extinct in the wild.

Siberian tigers were once common in the forests of Changbai Shan, and a few still roam the region. A favourite prey is wild boar. These ancestoral pigs live in large groups, probably for protection, and have an innate nervousness. A dense layer of subcutaneous fat helps insulate them against the freezing northeastern winters. Keen noses help them search out roots, fungi and dry grasses. Their favourite winter food is walnuts, and finding one nearly always results in a squabble. Small birds such as Eurasian nuthatches follow the boar through the forest, picking up any stray pieces of walnut that are dropped, for boar are notoriously messy eaters.

SIBERIAN TIGERS

today, Siberian tigers are, by comparison with the state of the other tiger subspecies, the most stable and secure, partly because the old-growth forests where they live have been relatively well preserved

The Siberian, or Amur, tiger is the largest of the five subspecies of tigers, with some males weighing up to 300kg (660 pounds). Originally found in the boreal forests of the Russian far northeast, the Korean peninsula and parts of northeast China, there are thought to be about 600 individuals left in the wild. Numbers dropped to about 40 in the last century because of heavy poaching, deforestation and hunting of their prey.

During the cold war, the tigers' habitat became off-limits to people, and their numbers began to recover. In the 1990s, Russia began to protect the tigers against renewed poaching, mainly for Chinese traditional medicine. Today, Siberian tigers are, by comparison with the state of the other tiger subspecies, the most stable and secure, partly because the old-growth forests where they live have been relatively well preserved. The south China tiger, though, is almost certainly extinct in the wild.

Breeding projects in China and in western zoos have raised many hundreds of Siberian tigers, many of which have been used for traditional medicine. But captive-bred tigers can't satisfy the demand for tiger bones and, conservationists say, fuel and mask the illegal trade. So poaching remains a threat to the wild population.

ice worlds

The threat from the people of the north was real. These tribes invaded and conquered the Han on numerous occasions. Today, many of these ethnic groups survive and carry on at least some of their traditions. One of the smallest, the Hezhe, live in the northeastern province of Heilongjiang and until recently were semi-nomadic. The Heilongjiang River – Black Dragon River – separates China and Siberia, where it is called the Amur. In midwinter, the river ice can be 80cm (31 inches) thick, and though it creaks and cracks, it is solid enough to make a means of getting around by

dog-sled or bicycle. But for those whose world involves fish, thick ice can be a problem.

The river harbours 225kg (496-pound) sturgeon, and one or two older Hezhe people still make clothes from their skin, though the practice is likely to disappear with them. In winter, catching fish is far from easy. The Hezhe make two holes in the ice 20 metres (66 feet) apart. Into one hole is lowered a string with a stone tied at the end to keep it taut. A long bamboo pole with a hook at the end is inserted into the second hole and used to hook the string. The fisherman then threads nets between the two holes – and waits. Catches of big fish are very rare now, and the fishermen usually make do with smaller salmon, trout and catfish.

ABOVE
Ewenki elder
Maliya Suo, at 81.
She is one of only
about 30 people in
China still herding
reindeer.

BELOW
An ice city built out of blocks of ice carved from
the Songhua River for the Harbin ice festival.
Ten thousand people spent 18 days creating it.

at dusk, the magic begins. The Harbin ice world
begins to glow from neon lights running
through each ice block

~

Far from struggling through the Siberian winter, the people of northeast China positively celebrate it. The world-famous Harbin ice festival attracts visitors from all over the world each January. Huge blocks of ice are carved out of the frozen Songhua River, and an extraordinary city of ice is built. It takes 10,000 people 18 days to construct the buildings, monuments and carvings out of 100,000 cubic metres (3.5 million cubic feet) of ice, some reaching a height of 80 metres (262 feet). Ice-carving competitions attract artists from a dozen countries.

At dusk, the magic begins. The Harbin ice world begins to glow from neon lights running through each ice block. The air fills with the smell of candied fruit and the screams of children as they shoot down giant ice slides, and a thousand camera flashes record the gaudy scene. Come springtime, the whole city simply melts back into the Songhua River.

In April, there is still plenty of snow on the ground near the border of Heilongjiang and Inner Mongolia. The nomadic Ewenki people are on the move again to new areas of forest. These are the only people in China to herd reindeer, and their animals need to eat large amounts of lichen. The lichen takes three to five years to recover, and so they have to keep moving. In spring, the reindeer also give

birth. Maliya Suo is in her eighties, but with her family, she takes a herd of about 400 animals deep into the forest. A large birthing enclosure is built, and as many of the pregnant reindeer as can be found are herded up into it. The Ewenki people rely on the reindeer to carry their tents, as well as for food and skins for clothing, and reindeer antlers are sold for traditional Chinese medicine. At birthing time, the reindeer also rely on the Ewenki – special wood wrapped in weeds is burnt to keep insects and mosquitoes away from the newborn calves.

the great grasslands

Northern China is covered by a huge area of grassland, much of it in Inner Mongolia. Spring comes late to this northern region, but when it does, a dazzling variety of wildflowers covers the land, including many that are familiar to us as garden plants, from purple iris and yellow, orange and red lilies to delicate pink and red roses. Scattered over the vast grasslands are important wetlands. Within Bayanbulak, in Xinjiang, is the biggest breeding ground in the world for whooper

swans. Tens of thousands of them migrate to the remote, safe wetland of Swan Lake Natural Reserve. Bayanbulak is also home for nomadic Mongolians. They revere the swans as divine birds and protect them against hunting and egg-collecting. The adult whoopers feed mainly on aquatic vegetation in the lake, but the cygnets need protein to build their flight muscles and so eat mainly aquatic insects and other invertebrates. A month after hatching, the weight of the cygnets will increase tenfold, and they will continue to grow for another three months or so before the onset of winter in October forces them to migrate south.

Inner Mongolia is the third largest province in China. In midsummer, when the grass is at its most lush, the Mongolians celebrate with the festival of Nadam. These people are famous warriors, and crowds of thousands gather to watch competitive archery and wrestling as well as some more modern games such as car-racing and even rugby. The sport they all excel at, though, is horse racing. It was the Mongolians' skill and tactics at conducting warfare on horseback that allowed Genghis Khan to establish the largest empire on Earth in the thirteenth century, and it is said that Mongolians are born in the saddle.

BELOW
Young riders taking part in the Nadam horse races, the biggest event in the Mongolian year.

Children as young as five compete in the Nadam races. As you walk past competitors preparing for the big race, you sense anxiety. There's no proper starting line in this vast landscape, and in chaotic fashion, before anyone seems ready, someone yells 'Go!' Hundreds of riders charge across the billowing grasslands for 30km (19 miles) or more, over undulating hills and streams – the small but powerful

XANADU

The Golden Lotuses Plain of Inner Mongolia lies only 275km (170 miles) north of Beijing, and yet it feels a world away. Mongolian yurts dot the gently undulating landscape, and the wind blows gently through the tall grasses and wildflowers. But in one place, the hills take on a strange form. Grass-covered ruined walls of an ancient city run in a large rectangle for 9km (5.5 miles). This is Yuan Shangdu, the summer palace of Kublai Khan – leader of China in the late thirteenth century. Known in the west as Xanadu, it was made famous by Samuel Coleridge's poem:

> In Xanadu did Kubla Khan
> A stately pleasure-dome decree.

As you walk around Xanadu, there is little evidence of this once mighty encampment. Pieces of green and black pottery, the remains of roof-tiles, push to the surface through the wildflowers. In the centre of the walled city are vestiges of a small room. It was here that Kublai Khan first greeted the traveller Marco Polo in 1275. The journey to reach Xanadu from Venice took 4 years, and Marco Polo spent another 17 years learning the ways of the Mongol court, before returning to regale Europe with his tales of the Orient. He reached home in 1295, and by 1368, the Mongol empire was over and the great Ming dynasty had begun.

as you walk around Xanadu, little evidence remains that this was once a mighty encampment

Mongolian horses truly in their element. Crowds gather at the finish-line, willing the riders home. It's a great honour to be crowned the best horseman, and the winning horse will be given the title 'forehead of ten thousand racehorses'. There are many tears shed by the young runners-up, with something special in store for the child in last place – he or she is paraded in front of the whole crowd, which shouts encouragement for greater confidence and a better result next time.

the waterless place

The western end of Inner Mongolia is very different in character from the east. The lush grasses which characterize the Mongolian steppe begin to disappear the further west you travel, and hardier plants and shrubs take over. This part of China forms the southern end of the great Gobi Desert, where conditions for survival are as

this part of China forms the southern end of the great Gobi Desert,
where conditions for survival are as extreme as can be found
anywhere. Outside the north and south poles, it's the least inhabited
place on Earth — in the world's most populous country.

~

extreme as can be found anywhere. Outside the north and south poles, it's the least inhabited place on Earth – in the world's most populous country.

Within the Gobi, which means waterless place in Mongolian, is a unique desert region called Badain Jaran, rarely visited, being a seven-day camel ride from the nearest town – itself absurdly remote. The Badain Jaran is China's third-largest desert (49,000 square kilometres – 18,920 square miles), and hidden within the desert's heart are what are now considered to be the world's tallest sand dunes, as high as 528 metres (1732 feet) tall. Walking over the dunes here is difficult, each footstep causing you to slip and sink. But a few feet below the surface of these mega-dunes is stability of a sort. Each year, a few salt-tolerant desert plants manage to gain a slight grip, fastening the sand particles together and giving still more plants a foothold. The following year, new mini-dunes appear over the top, are stabilized and then are lost into the heart of the mega-dune. Some Chinese scientists believe the centres of the mega-dunes could be tens of millions of years old.

BELOW
The Badain Jaran sand dunes, the tallest stationary dunes in the world – some more than 500 metres (1640 feet) high. Between the mega-dunes lie more than 100 lakes, supplied with water from an underground reservoir that dates back before the last ice age.

West of Badain Jaran, the ground is barren and rocky, and a deathly silence fills the hot, dry air. The few people who do live here speak of strange goings-on and bizarre creatures, none more so than the allghoi khorkhoi, or Mongolian death worm. Rumoured to be more than half a metre (2 feet) long and blood-red in colour, the death worm is said to be able to send out bursts of electric current to shock its prey and to spit venom that burns its victims. There's no evidence that the worm exists, but the locals swear they have seen it, though talking about it is considered very bad luck.

The ground here is peppered with thousands and thousands of holes, but the creatures who live in them are harmless. Semi-deserts such as the Gobi are often populated by great numbers of rodents. During the day, it is far too hot to be above ground – temperatures here reach 45°C (113°F) in the summer – but at night, the Gobi is a flurry of activity. Hamsters emerge into the moonlight – at least five different species are found here – and scurry around in search of the scarce food supplies. They may cover the human equivalent of four marathons every night. When the hamsters do find food, it's often a bonanza, such as a harvest of grass seeds that can sustain them for weeks. Their large elastic cheek pouches enable them to carry huge amounts back to their burrows, where a hamster may store hundreds of times its body-weight in food. At first light, the dunes will be crisscrossed with thousands of tiny paw-prints revealing the night's activity. This is the time when saker falcons and long legged buzzards will be patrolling over the dunes looking for stragglers returning home late.

Hamsters will normally live only for two years, and so these little creatures are in a big hurry to reproduce. Some desert species of hamster give birth 16 days after mating – one of the shortest pregnancies of any mammal. In fact, some hamsters have been known to wean a litter, give birth to a second litter and get pregnant with a third all on the same day.

the end of the wall

A long, meandering line of rubble in the southern Gobi Desert marks the path of the Great Wall. It has tumbled to the ground, but the climate is too dry for the mud-brick to disintegrate back into the parched earth. Following the remains of the wall west from Inner Mongolia, you enter the province of Gansu. Here the wall comes to an abrupt end as you arrive at the great fortress of Jiayuguan. During the Ming dynasty, Jiayuguan was legendary as the western frontier post where criminals were brought to be exiled into the deserts beyond, their final laments scrawled on the outer doors of the fortress. Construction of the fort was said to have been so meticulously planned that only one brick out of one hundred thousand was left over after the structure was complete. Beyond this far-western point, a barrier was considered unnecessary – the desert being the perfect natural one.

Jiayuguan prevented many foreign hordes from entering China, but it also took on another important function. It was a key stopping point along the legendary Silk Road that joined Persia and even Europe with China as long ago as the Roman Empire. Fabulous wealth was traded along its path, and great fortunes were made – but at considerable risk. The desert outside of Jiayuguan was so formidable that traders could cross it only with the assistance of a very special animal.

About 4500 years ago, Bactrian camels were probably domesticated in Bactria, in ancient Persia (Iran) and became the most important means of travelling and transporting goods across the deserts of Central Asia. Capable of carrying loads of almost 300kg (661 pounds), for 50km (31 miles) every day, long trains of these camels made large-scale trade a possibility along the Silk Road. Camels are adapted to desert life in a number of ways. They have wide, flattened feet to prevent them sinking into sand, long eyelashes to stop wind-borne sand getting into their eyes and nostrils that they can seal. But it is their ability to survive several days without drinking that sets them apart from any other pack animals. When a camel arrives at a watering hole, it can drink almost 60 litres (16 gallons) at a time. Travel by camel, though, is far from comfortable. Often they stand motionless, refusing to budge. Or they move suddenly, throwing you forwards; and if they don't like you, they might bite or spit. Crossing the deserts of northwest China by camel would have involved several years of hardship and discomfort.

The Silk Road leaves Jiayuguan and enters the vast province of Xinjiang. Very soon you notice a difference. Xinjiang is a place of extremes – occupying one sixth of China, 1.6 million square kilometres (61,776 square miles), it has eight international borders (Russia, Mongolia, Kazakhstan, Kyrgyzstan, Tajikistan, Afghanistan, Pakistan and India), as well as four provincial borders (Qinghai, Gansu, Inner Mongolia and Tibet). Xinjiang boasts the second highest mountain on Earth – K2, or Qiaogeli, as it is known to the Chinese – which stands 8611metres (28,251 feet) high on the border with Pakistan, and the second lowest place on Earth – Aydingkol Lake, in the Turpan Basin, at 154 metres (505 feet) below sea level.

Much of the province is even drier than the Gobi. The capital city of Ürümqi is the furthest city on Earth from any ocean (2250 kilometres – 1398 miles). But it is the people of this province that suggest that this is not the familiar China. Most are Muslims, speak Arabic, Persian or Turkic languages, and look more Middle Eastern than Chinese. There are at least 14 different ethnic groups living in Xinjiang, including Kazakhs, Kyrgyz, Uzbeks, Tajiks, Russians, Tartars, Mongols, Hui, Han, Xibo, Manchus, Dahours, Tibetans and, the largest group, the Uyghurs.

The Uyghurs have a legend that they are descended from the union between a Turkic boy and a she-wolf. Originally shamanistic, they then became Buddhists before converting to Islam in the tenth century. For a long time the Uyghurs controlled the Silk Road here, knowing the remotest corners and trade routes through this province, but there are places where even they still fear to go.

OVERLEAF
Shifting dunes in China's largest desert, the Taklimakan. There were once thriving cities here. Today, the region supplies China with oil.

the desert of death

The Taklimakan has been translated as, 'you go in and never come out again'. Up to 80 per cent is covered by shifting sand dunes, which can move 20 metres (66 feet) in a year – the second largest shifting sand desert on Earth, covering 337,600 square kilometres (130,348 square miles). A changing landscape can be more dangerous to travellers than the heat and lack of water, for if you lose your way, you cannot retrace your steps. Marco Polo remarked while crossing the Taklimakan:

> Sometimes the stray trader will hear as it were the tramp and hum of a great cavalcade of people away from the real line of march, and taking this to be their own company they will follow the sound; and when day breaks they find that a cheat has been put on them and that they are in an ill plight. Even in daytime one hears those spirits talking. And sometimes you shall hear the sound of a variety of musical instruments.

As winds whip across the dunes, you can sometimes hear strange groaning sounds, the origin of which has baffled scientists and travellers alike. The desert has not only claimed many lives but also claimed back many of the ancient Silk Road cities that were once scattered across it. Here and there the wooden beams of buildings emerge from the sand, only to disappear again. No wonder this desert has been called 'cemetery of civilizations', 'sea of death', 'fury of God'.

The Taklimakan, however, is surprisingly rich in wildlife. A number of very long rivers course under and through the dunes, bringing water from distant mountains. Where there is water, plants will find a way to grow, and this region is famous for poplar trees growing among the dunes. At least 93 mammal species, 25 reptile species and even 3 amphibian species are found in the Taklimakan, but the most numerous species are birds – 290 different kinds.

Birds can survive here because they can fly to new feeding grounds when conditions get too tough. But the rare Xinjiang ground jay – a small, ground-hopping bird with large, strong talons and a long, decurved bill – caches food for when times get really tough. How it remembers where the caches are, given that the sand dunes are constantly moving, remains a mystery. But if it has brains, it has brawn, too. With its powerful feet, it will dig down in search of roots and runners, yanking them out with its bill.

People have also found ways of surviving here, and there is evidence of human activity in the Taklimakan for more than 10,000 years. The early Silk Road pioneers established trade routes and cities more than 2000 years ago, but these were virtually abandoned by the eighth century in favour of travel by sea. Marco Polo's journey in the thirteenth century took him past cities already deserted for hundreds of years. The nineteenth century saw a flurry of explorers in search of the ancient Silk Road cultures. Today, oil exploration is the main activity, and this desert is the key domestic supplier of oil to the world's fastest growing economy. Conditions may be difficult, but there have always been resourceful people willing to brave them. In fact, 'you go in and never come out again' is a mistranslation of the name Taklimakan, ascribed to Sven Hedin, a Swedish explorer of the nineteenth century. In the Uyghur language, it actually means 'old home'.

Southeast of the Taklimakan, the dunes disappear and the ground is hard and flat. This is Lop Nur, one of the most remote corners of the planet. A Jin dynasty monk called Fa Xian wrote of Lop Nur in his book *A Record of Buddhist Kingdoms*:

DRAGON BONES

Northern China has been the site of many of the most important and spectacular palaeontological finds of all time. Bones of dinosaurs, called kong-long (terrible dragons), have been known about in China for more than 2000 years and may have led to dragon mythology.

In 1921, teeth found at Zhoukoudian, near Beijing, were ascribed to 'Peking Man'. In 1929 an almost complete skull cap was found of *Homo erectus*. This was not the first *H. erectus* fossil find, but Zhoukoudian has become the best studied site and helped to determine this species as a true intermediary between our ape ancestors and modern humans.

In 1922, the American explorer Roy Chapman Andrews set off from Beijing on his expedition to Outer Mongolia, where he discovered the world's first dinosaur egg fossils as well as the first *Velociraptor*. He was reputed to be the inspiration for the character Indiana Jones and had many scrapes with death on his travels.

The Yixian rock formation in northeastern Liaoning province has been described as the world's best fossil site. Not only has it unearthed *Sinosauropteryx*, the most important bird ancestor fossil since *Archaeopteryx*, but fossils from almost every animal group can be found here, some with almost perfect detail. Most of the creatures unearthed were buried in volcanic ash about 125 million years ago. To top it all, the world's oldest known flower fossils have also been found here.

The discovery of a winged-mammal fossil in Inner Mongolia suggests that mammals may have developed gliding flight before even early birds took to the skies. *Volaticotherium antiquus*, meaning 'ancient gliding beast', has a skin membrane covered in fur and, like the 'proto-birds', lived at least 125 million years ago.

In 2001, more than 100 of the largest dinosaur footprints were found in Gansu province. Each measures more than 1.5 metres (nearly 5 feet) in length, and they are all separated by strides 3.75 metres (12.3 feet) long. This suggests a dinosaur that was at least 20 metres (66 feet) long, weighing about 50 tons. It lived about 170 million years ago.

The oldest known 'modern' bird was found in Gansu province in 1983. Named *Gansus yumenensis*, it had webbed feet and lived 110 million years ago.

the Yixian rock formation in northeastern Liaoning province has been described as the world's best fossil site. Not only has it unearthed Sinosauropteryx, the most important bird ancestor fossil since Archaeopteryx, but fossils from almost every animal group can be found here, some with almost perfect detail

Evil spirits and hot winds predominate in the Sand River, those who encounter them will surely die. There are no birds in the sky nor animals on land ... only the desiccated bones of the dead mark the way travellers had taken.

Lop Nur is extremely dry and dusty, blasted by strong winds that howl through from the northwest. Over millions of years, they have helped carve giant structures called yardangs (meaning precipitous mud hill in the Uyghur language) arranged along the pathways of prevailing winds. But on closer inspection, there is evidence that these bizarre structures have been created partly by the action of fast-flowing water. Satellite images of Lop Nur show many concentric rings around the lowest point (dubbed 'the ear', because of its unusual shape), indicating that a lake was once here.

BELOW
Wild Bactrian camels in the Gobi Desert. Only about 650 remain in the desert regions of China (and 350 or so in Mongolia), which makes Bactrian camels rarer than wild giant pandas.

The *Book of Mountains and Seas,* written between 770 BC and 221 BC, records Lop Lake as a vast body of water. This great inland lake was formed about 2 million years ago, but tilting caused by tectonic movement drained large parts of it. Over the years,

it continued to grow and shrink as shifting sands changed the course of the desert rivers. But in the twentieth century, humans diverted much of the Taklimakan's rivers for their own uses, and in 1972, Lop Lake finally dried up. You can still see fishing boats sticking out of the ground – ghostly reminders of how precarious desert life can be. The surface of Lop Nur is covered in a very fine dust of clay and eroded mollusc shells. Strong winds whipping across the lake pick up the dust and

ABOVE
A field of yardangs (here in Yumenguan in Gansu province) – ridges created by sand-carrying winds that always blow in the same direction.

send it skywards, carrying it for thousands of miles across northern China as far as Beijing, nearly 2200km (1367 miles) from Lop Nur, and even on to the west coast of North America.

The deserts of Xinjiang are so far from the sea that water vapour simply can't reach them. The nearest source of water is the monsoon on the Indian subcontinent, but the Himalayas and the Tibetan Plateau, which form the southern boundary of

the Taklimakan, block almost all moisture. This hell-baked hinterland is one of the last refuges of one of the world's most endangered animals – the wild Bactrian camel. It's now thought that there are fewer than 1000 wild Bactrians left, scattered across Lop Nur and parts of the Gobi Desert.

Groups of wild camels migrate along well-known pathways through the deserts, as far from human settlements as possible. Their paths need to take them near water sources, but like their domestic cousins, they can go for many days without food or water, and are well suited to survival in harsh, dry, scorching conditions. Bactrian camels have been seen drinking from the brackish pools that characterize the deserts of western China. No other mammal could survive doing this. It's also hard to believe that any animal that lives on the desiccated stumps of dry desert plants and tumbleweeds could grow to 2.3 metres (7.5 feet) tall and weigh 450kg (992 pounds). But these are resourceful animals and will even gnaw the sun-bleached bones of their ancestors to obtain minerals. Though wild Bactrian camels are protected by Chinese law, they are under severe threat from poaching and loss of habitat. Wariness of humans and the harshness of the deserts where they live might be their salvation, but it also makes them nearly impossible for scientists to study and conserve.

the northern silk road

The route taken by ancient traders through the Taklimakan is known as the southern Silk Road, but it was often considered too perilous, and caravans of traders would instead take the northern Silk Road, which skirted the top of the desert. Conditions here are less harsh and water sources more reliable, but bandits made travelling along this route even more risky than the southern one. Leaving Jiayuguan Fortress, this time heading northwest, one of the first places you encounter is Dunhuang, a small town with a big surprise.

Of the many valuable imports that came from the west along the northern Silk Road and through this ancient town, there was one that was not a commodity – Buddhism. Nobody knows for certain when Buddhism entered China, but in the steep sided valley of Mogao, just south of Dunhuang, is a line of cliffs that hide the world's most important Buddhist sculptures. The earliest Buddhist iconography here dates from about 336 AD, showing the Buddha appearing to a local monk as a thousand points of light. Nearly 500 caves in this World Heritage Site contain painted murals, covering about 37,160 square metres (400,000 square feet) of wall and spanning a thousand years. Though more than 45,000 murals are still in place, many of the most important ones, as well as tens of thousands of manuscripts and some of the world's first printed books, were plundered and taken to the west over the past 130 years.

West of Dunhuang, the road leaves Gansu province and eventually runs into the formidable Flaming Mountains – part of the Tian Shan in Xinjiang province. This 100km-long (62-mile) range was immortalized in the sixteenth-century tale *Journey to the West*, where the monk Xuan Zang and his companions Pigsy, Monkey and Sandy travel to India to fetch Buddhist scriptures. In crossing the Flaming Mountains, Monkey's backside gets scorched, which, according to local legend, is why monkeys often have red bottoms.

Not far from here is Aydingkol Lake, in the Turpan Basin, the hottest place in China and the second lowest place on Earth at 154 metres (505 feet) below sea level. The salty substrate is like quicksand, and heat haze causes mirages that confuse travellers. Delirium quickly sets in as temperatures soar to 50°C (122°F) degrees, and the ground temperature has been recorded at nearly 80°C (176°F). Standing at Aydingkol Lake is like being in front of an open oven. When the wind blows, it feels even hotter, as more heat is brought in from the surrounding deserts. Nothing stirs,

OVERLEAF, BOTTOM
The caves at Mogao, near Dunhuang, on the Silk Road. Buddhism entered China along the Silk Road, and these 492 cave cells and sanctuaries are famous for their wall paintings and statues, spanning 1000 years of Buddhist art.

OVERLEAF, TOP
A Mogao cave painting depicting the 'Western Paradise of the Pure Land'.

SECRETS OF SILK

today, silk is manufactured in China using modern technology, but along certain remote parts of the ancient Silk Road such as Hotan in Xinjiang province, the method of production hasn't changed for thousands of years

Silk is a long strand of protein spun into a cocoon by larvae of the silkworm moth *Bombyx mori*. Legend has it that, more than 5000 years ago, a silkworm cocoon fell out of a tree and into the teacup of the Chinese Princess Xi Lingshi. From the strand that unwound in her tea, the first silk garments were woven. The lustrous quality of silk rapidly became famous, and by 1070 BC, an Egyptian mummy was buried swaddled in Chinese silk. At the height of the Roman empire, silk was so commonly worn that the Emperor passed a decree associating silk with decadence and banning it, but this was largely ignored. The secret of silk manufacture was carefully guarded, and this, coupled with the huge demand, helped open up a trade route across Asia – the Silk Road. Historically, silk production in China was based further south in the heartland. By about AD 300, silk production had been established in India with silkworms rumoured to have been smuggled out of China for huge financial rewards. Today, silk is manufactured in China using modern technology, but along certain remote parts of the ancient Silk Road such as Hotan in Xinjiang province, the method of production hasn't changed for thousands of years.

Each silkworm lays several hundred eggs, which hatch into tiny caterpillars. These are fed on mulberry leaves and increase in weight 10,000 times in just 50 days. About a quarter of a silkworm's body is made up of silk glands, used to spin a cocoon from a single strand of silk up to 5km (3 miles) long. Before the adult moths can emerge, the cocoons are thrown into boiling water. The unravelling strands of silk are gathered up and spun into thread and wound around bobbins, to be woven into sheets of fabric.

since nothing can live here. Yet only 55km (34 miles) northwest of the lake is the oasis town of Turpan – green, leafy and famous throughout China for its grapes. Almost every street is lined with trellises supporting the thirsty vines.

Their succour comes from vast networks of underground water channels, known as karez, that extend for thousands of kilometres all over this region. Entirely man-made, they start around the bases of distant mountains, carrying the water by

gravity to desert towns, often across enormous distances, with little water loss by evaporation. The karez required constant maintenance to keep the water flowing, and whole towns and civilizations depended on them. In the past few decades, modern pumps have enabled increased usage of water, and it may not be long before water no longer flows through the karez of Turpan.

The westernmost point on the Silk Road, before it leaves China for Central Asia, is the city of Kashgar, where both the northern and southern routes finally join back up. The city is rapidly modernizing and becoming more Chinese, but there is no mistaking the fact that this is an Islamic town. The focal point is the Id Kah Mosque, where Muslims from many different ethnic groups gather to worship. The markets are full of produce and wares from all over China as well as Central Asian countries, some of which haven't changed in thousands of years. Dried fruits and nuts, silk cloth, hand-woven carpets, fragrant woods, incense and precious stones, horses and goats now vie for space in the market stalls with music DVDs, baseball caps and computer software. This market town has always been a hub between many countries and cultures. Marco Polo's caravan would have stopped here to refuel and trade the pack-horses or yaks used in crossing the high Pamir Mountains for camels

ABOVE
A camel train carrying the tents and worldly goods of nomadic Kazakhs on their way down the Tian Shan to new pasture, where their sheep can overwinter.

needed to cross the deserts. His guides would also have been hired here, ancestors of the people still plying their trade in the bazaars nearly 750 years later.

Kashgar lies at the base of a number of important mountain ranges, but it is the Tian Shan, or Celestial Mountains, which have had the greatest impact on the geography and wildlife of Xinjiang province. A towering wall of rock 300km (186 miles) wide and running 1900km (1180 miles) east to west divides Xinjiang into two different regions. To the north, the slopes of the Tian Shan offer lush mountain pastures. Here Kazakh nomads – skilled horse-riders like all the nomadic people of northern China – fatten up their large flocks of sheep and goats in the short summer months. But in winter, they take their livestock down the Tian Shan and into the heart of China's second largest desert, the Gurbantünggüt, which lies in the Junggar Basin – the westernmost section of the Gobi Desert.

the junggar basin

What sets the Gurbantünggüt apart from other arid regions is the surprising amount of plantlife. Though the land is totally unsuitable for agriculture, more than 100 species of plant have been recorded here, and in places the ground is thick with saxaul – trees that can store water in their thick bark. Many animals graze on the flourish of vegetation here, including goitered gazelles, Mongolian wild asses, or khulans, and introduced Przewalski's wild horses. They are particularly adept at getting nutrients from the tough, desert-adapted plants, and the goitered gazelles rarely if ever drink, getting all their moisture from grazing alone. During the hottest part of the day, the gazelles and asses seek shelter if they can find any, or simply stand still to avoid overheating, becoming active when the cool evening descends.

In the summer, though, night is hardly any cooler – heat absorbed by the black rock of the desert is radiated back out. But many small creatures choose to emerge after dark, among them the animals with possibly the world's biggest feet relative to body size. There are at least five species of jerboa here, with feet up to four times as long as their front limbs. The foot bones are often fused to form a canon-bone to help them jump up to 3 metres (10 feet) in a single bound. Hair on the soles of their feet act as sand-shoes, enabling them to hop along without sinking into the sand. Long-eared jerboas are the most peculiar looking of all, with huge ears for homing in on the scurry of small insects.

This desert region covers an area of about 117,500 square kilometres (45,370 square miles) and is bounded on the north by the Altai Mountains, and on the south by the Tian Shan . The Altai is not nearly as formidable as its southern counterpart, and gaps in the range let cold, moist air from Siberia blow into the Junggar Basin. As temperatures drop in late autumn, snow begins to fall in the desert. This is one of the few places where snow and sand dunes mingle. Most animals are long gone by this

time, having hibernated deep underground or flown south to warmer climes. Temperatures can be bitterly cold, but the snow that blankets the desert will eventually soak into the ground to become the water that sustains the plants and animals next year. The Tian Shan to the south act as a barrier to this moisture leaving the basin, keeping the Taklimakan Desert parched and the Junggar Basin quenched.

During the winter, the Kazakh nomads come down from the mountains to graze their livestock in the snow-covered desert. This is far from ideal, but the high-altitude summer pastures are waist-deep in snow, and few animals could survive the winter there. The sheep and goats can find just enough food by scraping away the snow with

ABOVE
The Five Coloured Mountains of the Junggar Basin. The colours are layers of sediment deposited in the Jurassic and Cretaceous periods, either when the region was wet (grey or green layers) or arid (red layers).

their hooves and uncovering the plants. The Przewalski's horses are also digging for food and have to compete with the Kazakhs' livestock. In summer, they often seek out exposed hills and ridges to catch cool breezes, but in winter, their coats become thicker and darker to absorb more heat from the sun, and they seek out gullies and shelter from the wind, huddling together to share what little warmth they can generate.

This may be a difficult time for the wild animals of China's northern deserts, but it is when the Kazakhs go hunting. For more than 6000 years, falconry has been an important part of the culture of these nomadic people. Young birds of prey are taken from the wild and trained to hunt corsac foxes and hares in the hills around

how long the traditional practices of the nomads of northern China will continue is impossible to tell, but the fact that many have continued into the twenty-first century is an indication of the nomadic free spirit

their encampments. Ziya is 82 years old but still manages to go hunting in the snow each year. He dons his fox-fur-lined coat and hat to protect himself against bitter winds and wears a thick leather glove to stop his bird's talons from ripping his flesh. A golden eagle weighing nearly 7kg (15 pounds) and wearing a leather hood is hoisted onto his arm, and they ride off to the hunting grounds on horseback. Reaching a high point in the landscape, Ziya and his eagle will look for movement far below. The eagle has eyesight far more accurate than Ziya's. When something catches their attention, the eagle is loosed and dives at speeds of up to 130kph (80 mph), often killing its prey instantly. Before the bird dismembers the kill, Ziya will call it back to his arm with a cry of 'Ah ah!' and the wave of a fresh rabbit foot. This stops the pelt of the dead animal being damaged.

BELOW
A golden eagle patrolling the Taklimakan hill slopes for pikas, rabbits or other prey. Some Kazakhs still use wild-caught eagles for hunting game but eventually release their birds back into the wild.

When Ziya began hunting many years ago, he almost always returned with a fox or two, but these days, there is much less wildlife on the plains of northern China. Decades of overgrazing by livestock has degraded these rangelands. But though his eagle rarely catches anything any more, Ziya wants to pass on the skill of falconry to his grandchildren and great-grandchildren. A local saying pronounces that, 'the Kazakhs' wings are made of the best horses and the fiercest eagles.' Ziya's bird is now five years old, and he will keep it for another five years or so before releasing it back into the wild, where it can probably live for another 20 years.

The Kazakhs have always released their birds, recognizing that nature needs to regenerate itself if they are to carry on living in within it and from it. It is now illegal in China to hunt with birds of prey, as many of them are endangered and protected

PRZEWALSKI'S WILD HORSE

This small, sturdy horse, first described for western science by the Russian Nicolai Przewalski in 1881, was the last species of wild horse to be discovered and was last seen in the wild in 1969. It stands about 12 hands high and varies in colour from light brown in summer to reddish-brown in winter. It has a short, dark, bristly mane and a dark line running the length of its back. Huge herds once ranged across much of Central Asia, but encroaching flocks of domestic grazing animals on the steppes forced them into more and more marginal habitats, such as the Junggar Basin, where grazing is far from ideal. A number of Przewalski's horses were brought to Europe and the US in about 1900, and successful captive-breeding programmes kept the species alive. In 1992, the first Przewalski's horse reintroduction programme released 16 horses into what is now Hustai National Park in Outer Mongolia. China began breeding them in 1986, and 27 were released into Kalamaili Nature Reserve in the Junggar Basin in 2001. Several new foals are born in the wild each year, though conditions for survival are far from ideal in this part of China.

Przewalski's horses did range this far west into the semi-desert in the past, but increased competition with Kazakh livestock for scarce food is a concern, and when the nomadic Kazakhs pass through Kalamaili, there is a risk of their domestic horses mating with the wild horses. Though these animals evolved to endure winter on the Mongolian steppe, the zoo-bred animals have not yet adapted to the harsh conditions, and in winter, food is put out for them, as deep snow can cover what little food the desert has to offer.

China began breeding them in 1986, and 27 were released into Kalamaili Nature Reserve in the Junggar Basin in 2001. Several new foals are born in the wild each year

by law. But the nomadic people of northern China lead their lives according to the rhythm of the seasons and often don't speak any Mandarin Chinese and so may be unaware that this kind of hunting is illegal. How long the traditional practices of the nomads of northern China will continue is impossible to forecast, but the fact that many have continued into the twenty-first century is an indication of the nomadic free spirit.

BELOW
Przewalski's wild horses in their winter coats.
In harsh weather, they are given additional feed,
as these zoo-bred animals are not yet adapted
to the harsh conditions of the Junggar.

this small, sturdy horse, first described for western science by the
Russian Nicolai Przewalski in 1881, was the last species of
wild horse to be discovered

3

the tibetan plateau

AMONG THE MOST REMOTE AND INHOSPITABLE PLACES ON EARTH
are the Himalayas and the Tibetan Plateau. With temperatures of -40°C, snowstorms
in summer, boiling hot springs, salt-ridden lakes and oxygen-thin air, there appear
to be serious obstacles to life everywhere. Part is so inaccessible that it has even been
named the third pole. But it is also the most sacred and magical place on Earth for
millions of people.

Whether looked at from space or from the ground, the Himalayas and the
Tibetan Plateau are almost beyond comprehension in sheer scale. The Tibetan
Plateau is the highest expanse of land on the entire planet. It's a wilderness the size
of Europe and includes almost all of the world's territory higher than 4000 metres
(13,125 feet). Tibet is just a small part of it.

The Tibetan Plateau, as a geographical rather than political feature, extends
into neighbouring Qinghai and Sichuan provinces. At the southern rim of the
plateau, and inextricably linked to it, are the Himalayas. At 2900km (1800 miles)
long and with an average height of 5km (3 miles), these mountains are the true Great
Wall of China, visible from space. The Himalayan range is higher than anywhere else
on Earth, with hundreds of peaks more than 7000 metres (22,966 feet) high and 13
peaks higher than 8000 metres (26,247 feet), climaxing at Everest. It's an almost
insurmountable barricade to Nepal, Bhutan and India to the south.

Only recently have 'foreigners' been allowed to explore and study the area in
detail and to confirm theories about how the plateau came into existence. That fossil
sea creatures are found on Everest is the result of one of the most dramatic events in
the world's history – the head-on collision of two continents.

PREVIOUS PAGE
The Tibet western
region, with summer
storm clouds brewing
over the distant
Himalayas. On the
Tibetan Plateau, the
weather changes with
alarming rapidity.

ABOVE
A herd of female
kiangs, or Tibetan
wild asses. The herd
will be led to new
pastures by an old
female. In the
breeding season,
males will gather
together small
harems of females.

the great wind

A hundred million years ago, moving at a lightning-fast rate (geologically speaking) of 15cm (5.9 inches) per year, India barged and crushed its way 1932km (1200 miles) inland into Asia. The collision forced it downwards, and Asia was forced upwards, forging the highest mountains on Earth in the process. Millions of years later, the mountains continue to rise, at 3mm a year. The rocks themselves contain uranium, which makes the middle and lower sections unusually hot, expanding and forcing the plateau's upper crust upwards. It explains the flatness of the plateau itself – essentially, a giant crust floating on top of a 70km-deep (43-mile) bowl of once-molten rock.

Despite being so remote, this area has a significant effect on the Earth's climate. Like a great hotplate, the plateau heats up in the spring and summer, drawing up from over the Indian Ocean a great wind that carries with it a bellyful of water. An entire continent benefits from the annual monsoon rains that the wind brings with it, but the Tibetan Plateau itself sees almost none of it. As the wind starts to ride up the slopes of the Himalayas, it cools and the water vapour condenses out as rain, almost all of which has fallen by the time the wind gets to the peaks of the Himalayas. As a result, Nepal to the south of the mountains is lush and tropical, while Tibet in the rain-shadow of the mountains receives less than 46cm (18 inches) a year.

the pika grasslands

The Tibetan Plateau may at first appear barren, since much of its life is under the ground. If you look at any one section for long enough, you will eventually see a tiny flurry of dirt catapulted upwards, signalling the emergence of the small furry head of a Tibetan pika. This little relative of rabbits and hares is a keystone species of the plateau's ecology. The underground burrows of families provide homes for many other creatures, including ground jays and Tibetan snowfinches, which use them as shelter from the wind or even build nests in them, and reptiles, which use them as refuges, especially in the winter. The pika's constant excavations also break up and aerate the soil, which enables plants to put down roots. Without such a head start, few plants could survive here. The plants themselves are fuel for the grazers such as argali sheep – the largest sheep in the world, with great corkscrew horns – and kiangs, the wild asses that migrate over the plateau.

BELOW
A pika on the lookout for attack from the air as well as the ground.

BELOW
A shepherdess and her flock on the Tibetan Plateau in Qinghai province. Sheep and yaks provide the nomads of the region with transport, food, clothing, shelter and heating.

there are estimated to be 1.5 billion pikas consuming as much fresh grass each year as 20 million Tibetan sheep ... but without the furry farmers, there might not be much grassland in the first place

The pika itself is a very tasty morsel in a very short food chain – 90 per cent of pellets found under the nests of saker falcons contain pika remains, as do almost all the pellets of upland buzzards. Other predators include the square-faced Tibetan fox, which will sit patiently for ages waiting for a pika to poke its head above its burrow. A Tibetan brown bear trying to fatten up before hibernation doesn't have the time to sit and wait. Instead it tries to dig the pikas out of their burrows – though once out of its burrow, a pika can run rings around a lumbering bear.

With such a line-up of predators, a pika's life-expectancy is a mere 120 days. And today it has another enemy. Pikas consume grass, building tiny haystacks as food reserves close to their burrows. There are estimated to be 1.5 billion pikas consuming as much fresh grass each year as 20 million Tibetan sheep. Local farmers say they are destroying too much grassland on the plateau, and they try to exterminate them, but without the furry farmers, there might not be much grassland in the first place.

hotsprings and snakes

The evidence of the relentless activity continuing below the plateau is apparent on the surface – it's riddled with boiling springs that bubble, belch and spit clouds of sulphurous steam. Yet even such a challenging environment has been exploited. The hot-spring snake is found only in these Tibetan geothermal locations, at 4300 metres (14,107 feet) above sea level. An ancient relic from the plateau's low-lying past, before the collision with India, it holds the world record for reptilian high-altitude living and has survived the cold by squatting at these hot-spring locations.

LEFT
A harmless, inquisitive hot-spring snake – the world's highest-living snake – warming up before hunting.

OPPOSITE
A steaming hot spring – a reminder of the volcanic activity under the surface. The plateau is fractured by numerous hot springs, some of which are so hot that they spit like oil in a frying pan.

The snakes are harmless and even inquisitive, slithering close enough to humans to view them eye to eye. They are also gregarious, gathering in their dozens, not in the hot springs themselves but in and around the surrounding streams and rivers that are fed by the hot springs. Emerging mid-morning from their burrows, they slither across to a river or stream bank to warm up. Once warmed by the unfiltered sun, they slip into the shallow water. Here they wait, their heads perfectly still above the surface, occasionally twitching as they peer into the water. A group of three or four waiting side by side resemble jewel-like fishing floats. Every now and then, one will launch itself into the current, its sinuous body sidewinding at speed and then vanishing beneath the water, re-emerging downstream with a fish or a frog.

Thanks to the hot springs, about 50 snakes, each working a home range, can survive in an area of 600 square metres (6458 square feet), though in winter they are forced to hibernate underground. They seem to have few predators – perhaps the odd bird of prey – and the greatest danger they face is probably the hooves of yaks coming to the river to drink. But as more and more hot springs are developed for tourist attractions and bathing parks, the snakes face an uncertain future. In the warm-water area of one Buddhist monastery, however, these harmless snakes are protected, and their presence in the water is ignored by bathing monks.

the great yak

It's largely thanks to the yak that people have been able to exist up here on the 'roof of the world'. The yak is as central a part of the Tibetan Plateau as the camel is to the Sahara Desert, and as well adapted to its environment as is its desert counterpart. It provides people with transport and food – butter, milk, cheese, yogurt and meat

(dried yak meat can last for months). Its hide provides leather for boots, its coarse outer hair is used for making tents, and its soft inner hair is woven into warm blankets. Yak manure fertilizes fields and fuels fires. Even a yak's tail has a role to play in a land where nothing is wasted – used as a ceremonial object in Buddhist rituals. So widespread is the domestic yak that it is easy to take it for granted.

In the 1950s, an apparently brilliant plan was launched – cross-breeding lowland bulls with yaks to create mega-milk-producing hybrids. But in two years, most of the bulls were dead from altitude sickness. The dependable yak is all the more remarkable for its strength and agility – able to carry 70kg (154 pounds) over 5000-metre (16,405-foot) passes, surefooted enough to remain stable on treacherous tracks and able to wade through mountain torrents and barge waist-deep through snow. Anyone who has spent time at altitude will know how tiring just walking can be. Yet the yak is in its element – so much so that when moved to lower altitudes, anything below 3000 metres (9840 feet), it succumbs to diseases not encountered at higher altitudes.

The secret to the yak's success is vast lungs housed in an extended ribcage – 15 pairs of ribs compared to a cow's 13. Yaks also have three times the amount of red blood cells and a higher concentration of haemoglobin to oxygenate the blood. To help circulate the oxygen, yaks inhale deeply and rapidly, sounding (and acting) like miniature trains, and they grunt instead of moo – hence the local name 'grunting ox'.

ABOVE
A wild yak, one of relatively few small populations remaining in remote areas of the plateau. To protect it against the extreme cold of winter, a yak has a dense undercoat of fur and a dark brown, shaggy overcoat. Both sexes have huge horns, but those of the males are longest.

Impressive as the domestic yak is, it pales next to its wild relative. Standing 2 metres (6.6 feet) at the shoulder and weighing more than 800kg (1764 pounds), the native wild yak is formidable and has been known to kill domestic yaks. Living in herds of up to 200 in the remoter corners of the Changtang – the vast wilderness that extends through northern Tibet – wild yaks travel large distances across the alpine tundra in search of mosses and lichens. They are mostly black, but a tiny population of golden yaks lives in the Aru basin region of the Changtang.

All wild yaks are incredibly difficult to approach. On the vast open spaces, they can spy you from about 5km (3 miles) and then flee a further 20km (12 miles). But during the rut, when the bulls spar, they may stand their ground, raising their tails and waving them aggressively and even charging people who get too close.

Like many other creatures of the plateau, wild yaks were overhunted in the twentieth century, being the source of so many essential raw materials. Their demise was hastened with the arrival of vehicles, which made it easy for hunters to penetrate deep in wild yak country in winter, when the ground was frozen. Within 30 years, wild yaks in the southern part of the Changtang were virtually eliminated. Today, numbers have increased, possibly to 15,000 in various herds around the plateau. But that compares to about 12 million domestic yaks.

the tibetan unicorn

In the Middle Ages, intrepid European travellers brought back exquisite horns that inspired tales of a Tibetan unicorn. They were, in fact, from the chiru, or Tibetan antelope – perhaps even more extraordinary than the fabled unicorn. The chiru is not an antelope but related to the golden takin and the mountain goat of North America. The male is about a metre tall (just over 3 feet), and his head is crowned with two extremely long horns (72cm – 28 inches). Chiru spend all year on the Tibetan Plateau, travelling back and forth across the vast spaces, weathering the extreme temperatures and fierce winds.

Unfortunately, the adaptation that has helped them survive the harsh weather has caused their near extinction in the twentieth century. The chiru's coat has two layers: a dense outer one and a fine inner one with hair a fifth finer than human hair. The light, fine, soft 'shahtoosh', or 'king of wools', is the warmest wool in the world and is highly prized in fashion stores worldwide. It takes the death of five chirus to make just one shahtoosh shawl, which can sell for more than $15,000.

BELOW
Male chiru, displaying his long horns, with a female and calf. Chiru are incredibly wary of people, having been hunted nearly to extinction, not for their horns but for their soft, dense undercoat, regarded as the warmest wool in the world and made into hugely expensive shahtoosh shawls.

A century ago there were millions of chiru – a spectacle to rival those of the plains of East Africa – but such was the demand for shahtoosh that 20,000 chiru were killed annually. Though the trade was banned in 1975, demand continues, as the stakes are high and policing the chiru's remote habitat is a huge challenge.

In 1992, a Tibetan called Sonam Dhargye formed the Wild Yak Brigade – a group of volunteers who vowed to protect the chiru. They were successful until 1994, when the group tried to arrest 18 poachers carrying 2000 chiru skins. A fight broke out, and Sonam was killed. His brother-in-law Dakpa Dorjee became the new leader, and in four years the group rounded up a further 250 poachers. Then in November 1998, Dakpa was shot dead. He and Sonam became national heroes, and their tale became an acclaimed movie, *Kekexeli*. Between 1990 and 1998, Chinese agents seized 17,000 chiru pelts, 1100kg (2425 pounds) of chiru wool, 300 guns and 153 vehicles used by poachers. Today a special force, the Kekexeli District Protection Administration, has been drafted from former members of the Wild Yak Brigade.

It is still possible to find big groups of chiru, especially in winter, when the males gather around the females and begin their spectacular rut. The males' faces and legs grow darker, and once in their new coats, they look striking against the snow-capped landscape. The males fight to control groups of 20 or so females, locking horns in combat. Some die from their injuries, and the injured and weak are soon picked off by wolves or even snow leopards.

Following the rut, and driven by the need for food, the males migrate and don't meet up with the females and young again until the autumn, in preparation for the next rut. The females migrate up to 300km (186 miles), driven by the need to find a safe haven for their calves as much as for food. There are four separate populations of females, each using different calving grounds. The routes are not necessarily the most direct – they date from the ice age, passed down from generation to generation, the young ones learning from their mothers. The Qinghai-Tibet railway, completed in 2006, carved straight through one of these routes, but the authorities have built a bridge over it, and initial reports suggest the scheme is working. But it remains to be seen how both railway and chiru will fare in the long term.

For many years, the location of the calving grounds was a mystery – mostly due to the logistics of studying on the plateau. But in recent years, sites have been discovered in both Qinghai province and the Arjin Shan Lop Nur Nature Reserve. The female chiru choose summer breeding grounds where there are few potential predators on the calves – at this time of year wolves have their own cubs in the den and so cannot follow the chiru on their migration. Even so, the mortality of newborns is still high: a calf has a 50 per cent chance of dying before it's a month old, and only a third of calves will make it to two years old. Thanks to the efforts of the Wild Yak Brigade and others, and increasing awareness of the chiru's plight, the numbers are not dwindling at the rate they were. And so symbolic is the chiru of the Tibetan Plateau that it was chosen as one of the mascots for the 2008 Olympic Games.

sky burials

Vultures thrive on the plateau, where the thermals allow them to glide and soar effortlessly. Bearded vultures (lammergeiers), their wingspan nearly 3 metres (10 feet) wide, seem to defy gravity as they glide overhead, while smaller vultures such as Eurasian griffons circle in search of pickings. A flock of circling vultures in the sky may not, though, mean the presence of a dead yak below. The meal may be a sky burial laid out for them. Traditionally, when a person dies, the corpse is cut up and left for the vultures. Few non-Tibetans have witnessed this intensely private and sacred event, but it is widespread and intrinsic to Buddhist belief and Tibetan culture. The concept of rebirth after death means there is no need to keep the body, since the soul has already passed on.

Drigung Monastery is one of the most auspicious sites for sky burial in Tibet. The ritual begins with the monks chanting around the body and burning juniper. The body is then taken to a flat rock where it is cut into pieces by rogyapas, or body breakers. There is often laughter and chat as the body is hacked to pieces – but this is consistent with their belief that the body is merely an empty vessel. Meanwhile the circling vultures land nearby, awaiting their summons by the rogyapa. Though other cultures consider vultures dirty scavengers, the Tibetans respect the fact that they do not kill anything. The word for burial in Tibetan means 'giving alms to the birds' – an act of generosity reflecting the Tibetan concept of compassion for all beings. By doing this good deed, the donor of the body will accumulate merit and be reborn in better circumstances. On a more practical level, it's not feasible to bury or burn a corpse – the ground is too hard and wood is too precious. So sky burial is a perfect marriage of the practical and the spiritual.

snow leopards

There is no creature more symbolic of this high, remote landscape than the snow leopard. Yet it's one of the least known and seen – not surprising, considering the terrain it lives in. Between 3000 and 5400 metres (9840–17,720 feet), the snow-covered crags with their gullies, cliffs and rocky slopes might seem an impossible hunting area for a large predator, but the snow leopard thrives in these conditions. A formidable predator, measuring up to 2.3 metres (7.5 feet) long and weighing up to 54kg (119 pounds), it has short, stout forelimbs and long hind limbs to negotiate the unpredictable terrain. Its lightly spotted, camouflaged coat is very warm – the fur on

ABOVE
A bearded vulture circling on the lookout for carcasses, including human ones, whose bones it will scavenge, dropping them onto rocks to break them open. It has a tongue specialized to scoop out the marrow.

OPPOSITE
A snow leopard – icon of this high, remote region – her belly full of a blue sheep kill. Snow leopards are still persecuted by farmers, who see them as threats to their livestock.

its belly is nearly 12cm (5 inches) long. It has large snowshoe-like paws and a tail nearly as long as its body, which helps it balance as it hunts on uneven and shifting ground. The tail also doubles as a warm scarf, wrapping around the cat's body when it's resting. Like many large animals on the plateau, the snow leopard has an oversized chest and lungs to compensate for the thin air. A large nasal cavity also enables the cold air to be warmed as it is inhaled, and the cat makes strange puffing sounds as it does so.

It needs a large home range that is well stocked with prey – up to 1000 square kilometres (386 square miles) in some areas, but just 30 in others. Blue sheep and marmots are usual prey, often killed following a leap up to six times the cat's own body length. China may have the world's largest population of snow leopards, though estimates range from 2000 to 5000. The rugged Tibetan Plateau is ideal territory for them, with the highest densities around Kyirong valley in the south.

Though given 'first-class-protected' status in China, snow leopards are still hunted for their pelts and their bones – prized in Chinese medicine (especially now tiger bones are harder to procure) and worth hundreds of dollars. Snow leopards, especially females with young cubs, may also kill sheep in areas where their natural prey has been hunted out, and so they are killed by the farmers. In the past decade, projects have been initiated to help find a way to achieve a balance between the needs of wildlife and those of people, and some are now making use of Tibetan Buddhist traditions to help prevent the killing of snow leopards for fur. Conservation is certainly part of the ancient culture. Indeed, most monasteries have a sacred site where the surrounding area is set aside for wildlife and hunting is forbidden.

mother of the universe

In 1852, an Indian mathematician, Radhanath Sikdar, took measurements of a mighty mountain in the far distance known as Peak XV. He was awestruck by his findings – at 29,000 feet (8839 metres), it was the highest peak in the world. Unable to discover any local name for the peak, the British Surveyor General of India was later to name the mountain after his predecessor, Colonel George Everest. Mount Everest was, in fact, already known by many names – Devgiri, Devadurga and Qomolangma, which in Tibetan means 'Mother of the Universe'. In 2002, the Chinese launched a campaign to reinstate the name Qomolangma, which they argued had been marked on a Chinese map for 280 years.

Whatever its name, this iconic mountain has fascinated people all over the world, attracting climbers like a super-sized magnet. When in 1923, the British mountaineer George Mallory was asked exactly why he wanted to climb such a terrifying mountain, battered by constant winds in temperatures that plummet far below zero and where breathing was close to impossible without oxygen, he famously replied 'because it is there'.

This mountain remains an awesome place even for climbers today, with all the modern paraphernalia. The main window for climbing is April/May, when the wind speed is at its lowest before the monsoon. There are two main routes up Everest – the southeast from Nepal (used by Hillary during his successful attempt in 1953) and the northeast from Tibet. Mallory chose the Tibetan route for his ill-fated third attempt, which starts at Rongbuk Monastery, the highest in the world (as most things seem to be on the Tibetan Plateau). It is still debated whether Mallory actually reached the summit in 1924, but in 1999, his body was found preserved in the ice at 8040 metres (26,378 feet), frozen for 75 years. He is still there. Somewhere else on the mountain is his camera, also preserved in the ice, which could contain irrefutable evidence of whether he reached the summit or not. Unfortunately the mountain has claimed others since Mallory's tragic expedition – nearly 200 climbers, many of whom remain entombed in the ice, to be recovered only when the Khumbu and Rongbuk glaciers finally carry them down to the base camps below.

Whichever route is taken, above 7200 metres (23,625 feet), climbers enter the notorious 'death zone'. Here the oxygen becomes so thin that most climbers have to use supplementary oxygen. The mountain has been climbed without it – Reinhold Messner achieved this in 1980 – but this is a truly exceptional and dangerous feat.

JUMPING TIGER HUNTING FLY

they can jump 30 times their own body lengths and are such successful hunters that the Chinese call them 'fly tigers'

When in 1924 Mallory and his team reached 6700m (22,000 feet) on Everest – the highest any human had climbed – they found that something had beaten them to it. Darting about at their feet were jumping spiders. It was thought they had been blown there from the valleys below, since it was hard to imagine what they could be feeding on. Yet they were seen again at the same location on a 1954 expedition. In fact, this incredible spider, *Euophrys omnisupertes*, is the highest permanent resident on the planet. As soon as the clouds clear and the air heats up (since less solar energy is absorbed by the thin air, 33°C (91°F) is possible) the spiders emerge from their shelters under rocks and begin to hunt. Two of their eight eyes are huge and scour the slopes for the movement of wind-blown flies and springtails. Being fluid feeders, they must catch their prey alive. They can jump 30 times their own body lengths and are such successful hunters that the Chinese call them 'fly tigers'.

LEFT
The Rongbuk glacier at 5500 metres (18,045 feet), spawned from Everest. Like all the glaciers here, it is shrinking, almost certainly as a result of global warming.

The air (or lack of it) is a major problem for climbers' safety – the atmospheric pressure at the top is one third that at sea level, and so there is just one third of the oxygen available to breathe. High-altitude sickness is a risk from 3000 metres (9840 feet) upwards, leading in extreme cases to cerebral oedema and death.

In 1996, the effect of altitude was compounded when a freak event occurred – known as 'the day the sky fell on Everest'. A seemingly calm day suddenly turned stormy, just as two jet streams of fast-moving air were passing over the mountain. These pushed the stratosphere and the air pressure downwards – equivalent to raising the summit of Everest a further 500 metres (1640 feet), with a resulting drop in oxygen. Eight climbers died. But this has not deterred people attempting to climb the mountain, which today is measured at 8850 metres (29,035 feet) – 11 metres (35 feet) higher than Mr Sikdar's original observations in 1852.

super-geese

The name bar-headed goose, referring to the two bars of dark feathers around its head, gives no indication of the remarkable feat this bird undertakes. Riding the jet streams that race at 322kph (200mph) above Everest, flocks of them on their annual migration to winter south of the Himalayas fly far above the low-oxygen 'death zone' encountered by climbers. Come spring, they return again to breed on the Tibetan Plateau. They achieve this through a combination of strong, aerodynamic bodies and large wings that allow them to cruise at 80kph (50mph) – faster with a good Himalayan tailwind. As they fly, they warm up, and so their wings don't freeze. A network of air sacs in their lungs creates a counter-current that extracts oxygen from each breath, which is effectively inhaled twice. Their haemoglobin also absorbs oxygen quickly, and their capillaries are densely packed, so that the blood is rapidly delivered to the muscle fibres. Extra myoglobin in the muscles can store the oxygen longer – the reason goose meat is deep red.

The difference between the flight feats of bar-headed geese and other birds is breathtaking – they fly an average 7500 metres (24,610 feet) higher. This begs the question of why they push their bodies to extreme limits to fly back and forth to the Tibetan Plateau. Before the clash of tectonic plates that buckled the plateau and built the mountains, Tibet was probably a lush landscape and a breeding paradise, especially for animals fleeing the approaching monsoon in the south. The geese almost certainly continued to fly the same routes as the mountains grew beneath them, each generation learning the route from their elders.

Today, the land north of the Himalayas is vastly different, but from a goose perspective, the plateau is riddled with suitable breeding sites – jewel-like lakes, such as Manasarovar (the highest freshwater lake in the world), Serling Tso in the Changtang and the salty waters of Qinghai, China's biggest inland saltwater lake. When the bar-headed geese arrive in May, these sites provide an ample supply of plants,

BELOW
Bar-headed geese at Qinghai Lake, their main breeding site and China's biggest inland saltwater lake. Avian flu has decimated their numbers here.

SACRED CRANES

In the marshlands surrounding the lakes on the Tibetan Plateau, some of the country's largest and most charismatic birds can be seen in spring performing exquisite courtship dances. Pairs of black-necked cranes bow, jump and flap their wings, their heads thrown back over their shoulders as they point their beaks skywards and sing to each other. The cranes spend the winter at relatively lower altitudes around Lhasa, where they forage in farmers' fields for leftover barley, but in spring they return to the high marshes to breed. Parents can be seen parading aristocratically, softly calling to their newborn chicks, who follow close behind as they search for fish and frogs.

Seventy per cent of the world's population of black-necked cranes live and breed on the Tibetan Plateau, but because of the challenges of working here, it was the last species of crane to be described by scientists – even though they were already well known to the Tibetans. In the seventeenth century the sixth Dalai Lama wrote:

> White crane, lend me your wings,
> I go no farther than Lithang,
> And thence, return again.

Tibetans believed he was predicting the site of his own reincarnation. Sure enough, the seventh Dalai Lama was born in 1708 in Lithang. Not surprisingly the black-necked crane is a sacred animal, and anyone convicted of killing one is imprisoned.

seventy per cent of the world's population of black-necked cranes live and breed on the Tibetan Plateau ... it was the last species of crane to be described by scientists

crustaceans and spawning fish. For safety, thousands of the geese nest communally, often within pecking distance of each other. A mother lays three or four eggs, which hatch about thirty days later, when the parents defend them against predators such as Tibetan foxes, which have a bonanza at hatching time. But today there is a far more serious threat to the geese – avian flu. In recent years, thousands of geese have died, and the breeding lakes of Qinghai and Tibet have been in quarantine.

BELOW
A Tibetan in his yak-hide boat on the Yarlung –
Tibet's longest river. Born on Mt Kailash,
'centre of the world', it flows off the plateau,
transforming into the holy Brahmaputra.

four major rivers flow from this region, initially following the cardinal points of the compass. These are some of the longest and most important rivers in Asia

Spring and summer are short in Tibet, resulting in a flurry of activity from both animals and people. In some areas, groups of Tibetans may be seen scouring the grassland, bent over in deep concentration. Occasionally one will reach forward and carefully dig something out of the soil – an earthy brown, root-like thing – and deposit it in a collecting bag. Great care is taken not to break the 10cm-long (4-inch) 'root'. It's possible to dig up 40 of these in a day, the proceeds from which may amount to half the collector's annual income.

This is yatsa gunbu – the world's weirdest harvest. In Tibetan, the name means 'summer grass, winter worm'. It's an apt description. The winter worm is a moth caterpillar. It spends up to five years growing, eating as much as it can – mainly the roots of grasses and other plants – in preparation for its pupation and transformation into a white ghost moth. But at the time it is getting ready to pupate underground, strange protuberances emerge from its head and start growing out of the soil, rising up 5–10cm (2–4 inches). They are the fruiting bodies of the fungus *Cordyceps sinensis*, which mature to release millions of tiny spores that are blown across the grasslands. Their ultimate destination is the body of a ghost moth caterpillar. Once a fungal spore makes contact, it grows into its host and infects its brain, causing it to burrow into the soil, as if it was about to pupate, but remaining much closer to the surface than it would normally do.

Now a mesh of fungal hyphae starts to feed on the contents of the caterpillar until the entire body is emptied of nutrients. At this point, and if conditions are right, the fungus will start to grow a fruiting body. It emerges out of the horn above the eyes of the head of the caterpillar's mummified body until it projects above ground, ready to start the cycle again – infecting new caterpillar hosts from the body of one of their own kind. It is this fruiting body that alerts the collectors to the mummified riches below.

ABOVE AND LEFT A hunter of the fruiting bodies *Cordyceps,* which emerge from the carcasses of their caterpillar hosts. Just 5g of the fungus is equivalent to 50g of ginseng.

The Tibetans first noticed the properties of yatsa gunbu when their domestic yaks appeared to have more energy once they had grazed on *Cordyceps*. It has been used as a traditional medicinal remedy for thousands of years – though only by the very wealthy. In the second century BC, the emperor used it for endurance and longevity. It was bartered for tea and silk, and was worth more than four times its weight in silver. The magic formula is believed to strengthen the system, regain energy, combat exhaustion and even help in erectile dysfunction and premature ejaculation.

In recent times, *Cordyceps* has become more widely available, and when it was adopted by the Chinese sporting community in 1993, the effects were extraordinary: two female athletes broke the 10,000-metre, 3000-metre and 1500-metre world records, prompting a massive surge in interest. Tests revealed that *Cordyceps* made it easier for the athletes to breathe and lowered their blood pressure. Today there are plans to make *Cordyceps* more widely available through commercial harvesting – not good news for the caterpillars or the future of the harvest. Though the Tibetans trade in them, they also regard them as treasure from the earth spirits, consistent with their belief in the sacred nature of all things.

the most sacred mountain of all

The landscape and creatures of the plateau have played their part in shaping the ideas of Tibetan Buddhists, but the influence and magic of this place reaches far beyond the plateau. Hindus in India believe in a legendary place called Mount Meru – a giant mountain that towers above its neighbours. It has four faces, made of crystal, gold, ruby and lapis lazuli. From its peak flow four rivers that run to the four quarters of the world. Meru is not only the centre of the world but also the ultimate destination of souls, since Shiva, god of destruction and regeneration, resides at its peak. Is this a truly fabulous myth or more than that?

BELOW
The mythical centre of the world – Meru – appears to be a reality in the form of Mount Kailash, a great pyramid of rock and ice.

ABOVE
Pilgrims of many
faiths gathering at
the end of their
pilgrimage to the
foot of the sacred
mountain of Kailash
in western Tibet.

There is a mountain in Tibet that closely matches the description of Meru. This is Kailash, which rises head and shoulders above its neighbours, its peak reaching 6638 metres (21,778 feet). Its name means 'precious jewel of stones', and the sides of this great pyramid of rock and ice are carved into four almost perfectly symmetrical faces, so sharply hewn that they appear to be man-made. The four sides are roughly aligned to the points of the compass, and four major rivers flow from this region, initially following the cardinal points of the compass. These are some of the longest and most important rivers in Asia – the Indus, the Sutlej, the Yarlung and the Karnali (Ghaghara), a major tributary of the Ganges. No wonder Kailash is a sacred place.

It is venerated not only by Hindus but also by Jains, who believe the founder of their faith achieved nirvana here. Buddhists believe it is the home of Demchok, a manifestation of Shiva who represents supreme bliss. Bons (people who follow the ancient shamanistic religion of Tibet, which pre-dates Buddhism) believe it to be the seat of all spiritual power. Some think Kailash was sacred even before all these ancient religions began. In recognition of the mountain's extraordinary status, it has never been climbed – rumours of a planned Spanish attempt in 2001 were quickly dispelled following an international outcry.

ABOVE
The Saga Dawa
festival flagpole
festooned with flags
carrying prayers
and messages. The
raising of the pole is
accompanied by
great celebration.

Whichever faith one may or may not believe in, it is impossible not to be moved by this place. To reach it involves a hair-raising four-day drive from Lhasa to the wild western corner of Tibet. It is a road that carves through the theatrical backdrop of the Himalayas in the south and the foreboding and seemingly endless wilderness of the Changtang in the north. It is also strewn with the remains of lorries and land cruisers – no journey to Kailash should be undertaken lightly. But perhaps that is the point. This is one of the most remote and sacred places in the world. It is the destination of a lifetime.

For Tibetans, pilgrimage is a journey from ignorance to enlightenment, from materialistic preoccupations to a sense of the interconnected nature of life. The word for pilgrimage, *neykhor*, means 'to circle around a sacred place'. From 1959 to 1980, physical pilgrimage was not permitted, but since the reopening of China, a sea of pilgrims arrive at the foot of Kailash each year, not just from Tibet but also from the Chinese heartland, Nepal, India and further afield. Most pilgrims time their arrival for Saga Dawa, the most important event in the Kailash calendar – a festival that celebrates the birth, death and enlightenment of Buddha. It takes place at the full moon in the fourth lunar month of the Tibetan calendar – roughly mid-June.

Pilgrims gather at the foot of Kailash, where there is a 24-metre (79-feet) flagpole. It is held upright by great streams of ropes festooned with prayer flags that flap loudly as the wind blows the prayers to heaven. On the first day of the festival, the chief Buddhist monk – the lama – proceeds slowly around the pole accompanied by an ear-blasting chorus of drums, cymbals and horns. The lama stops at the foot of the pole and recites prayers and blessings. Once the pole has received a suitable amount of attention from the lama, it is lowered, slumping to the ground like a felled tree. Immediately, pilgrims rush in and tear off the old flags, the ones nearest the top being the most auspicious. Once the pole is stripped as clean as a carcass, new flags are attached, each brought by a pilgrim and imbued with prayers and messages.

On the second day, the newly decorated flagpole is ablaze with fresh colour. Following more music and blessings from the lama and amid much shouting and general excitement, the pole is slowly raised. It halts frequently – the raised pole must be straight or it will be a bad omen for Tibet. Thanks to some careful coordination by the teams responsible for hauling the flagpole upwards, it finally stands true, and an almighty cheer follows as the pole flutters in the mountain breeze like a beautiful tree in bloom. Buddhists, Bons, Jains and Hindus sing, dance and pray. More prayers, written on colourful pieces of paper called 'wind horses', are thrown into the air. The sky is filled with these dancing wind horses, which float and flutter upwards towards the peak of Kailash.

It's not hard to believe Shiva really is there waiting for them. The pilgrims proceed to circumnavigate the mountain – a 52km-long (32-mile) arduous route that tests both body and spirit. Those who are super-fit can achieve the route in just one day, but the really devout perform full-length prostrations, crawling forward on hands and knees just a few feet at a time. Most appear on the far side of the mountain four days later, ragged yet elated, since they have just erased the sins of a lifetime. The Saga Dawa festival has been celebrated for more than 1000 years. Here, at the believers' centre of the world, is a rare vision of harmony between different faiths.

The Yarlung Tsangpo River flows east from Kailash, searching for a route off the plateau but blocked by the wall of mountains. It is Tibet's longest river, flowing 2000km (1243 miles) before it meets Namcha Barwa, the highest mountain in the Himalayas' eastern section. Here at last the Himalayan wall is broken, and the river carves southwards in the most dramatic fashion through the Tsangpo Gorge. This is the deepest gorge in the world – at 5382 metres (17,657 feet) – and one of the longest, through which the Yarlung roars, descending rapidly and finally tumbling into India, where it emerges as the holy Brahmaputra.

At its northern edge, the Tibetan Plateau also gives birth to the Yellow River, which flows down to the north China plain – birthplace of Chinese civilization. Other rivers that spring from the plateau's glaciers and lakes include the Mekong, Salween and Yangtze – making ten major rivers in all, sustaining the lives of 47 per cent of the world's population. No wonder these celestial heights are so revered and regarded as so precious.

OPPOSITE
Rarely seen in clear weather, the Tsangpo Gorge is deep enough to swallow three Grand Canyons. Here at the Big Bend, the river escapes the wall of mountains and finally tumbles through lush, monsoon-fed valleys into India.

yunnan

IN A REMOTE FORESTED VALLEY IN THE SOUTHWEST OF CHINA, early morning mist curls around the trees and calls of distant birds echo across the canopy. As the sun rises, spears of light penetrate the dense complex of foliage, transforming the interior into glades of greens and causing wisps of moisture to rise. Suddenly a new sound joins the dawn chorus – one that makes the hairs rise. Starting slowly, the eerie, whooping calls of a family of Wuliangshan black-crested gibbons build to a crescendo. Every morning, from a high branch overlooking the Wuliangshan valleys, the family joins in a territorial taunt, echoed by other troops in the area.

To hear gibbons sing in the wild today requires many days of trekking to the remotest of forests. That these primates exist in China may seem surprising until you remember that the southwesterly province of Yunnan, the only place in China that they are still found, is in Southeast Asia and harbours the last remnants of some of the world's most ancient tropical forests.

PREVIOUS PAGE
The forested volcanic region of Tengchong in the western corner of Yunnan on the border with Burma. The remnant tropical forests harbour plants and animals found nowhere else.

BELOW
White-cheeked gibbons, the female (right) grooming her mate. This is one of five species of gibbons found in Yunnan, all rare or endangered and existing only in isolated patches of remnant forest.

> Yow! Yow! Night gibbons cry.
> Soft, soft, dawn mists mesh.
> Are their voices far? Nearby?
> Just see the mountains piled up high.
> Having liked the East Hills's song,
> I now await West Cliff's reply.
> SHEN YUEH (441–513)

Steep-sided, forested hills straddle Yunnan's southern border with tropical Laos and Vietnam. Small herds of Asian elephants cross back and forth, following trails that their ancestors have probably used for centuries. Along the eastern border, the dense forests merge with those of Burma. Moving northwards, the forests stretch over undulating hills into the highland territories. But while Yunnan's southern hills lie across the tropic of Cancer, its northern borders are part of the Himalayan massif – a barrage of frozen, impenetrable peaks more than 6000 metres (19,685 feet) high. Transected by immense rivers that run the length of Yunnan, the two worlds are as different as you can get. In fact, this comparatively small province, covering little more than than 4 per cent of China, is one of the most diverse places on the planet.

ABOVE
A glimpse of some of China's last wild elephants. Once elephants roamed across the country as far as Beijing, but they are now found only in the nature reserves of Yunnan close to the borders of Burma and Laos. The total population may be as low as 250.

snow monkeys

The snowy peak of the Tibetan sacred mountain of Kawa Karpo marks the northernmost territory of Yunnan. At nearly 7000 metres (22,970 feet) high, this pyramid of ice and snow is the crown of the Hengduan Shan. Tibetan winds whip across its alpine slopes and down into pine and larch forests below. In winter – November through until March – at altitudes above 3000 metres (9840 feet), conditions are bitterly cold but extremely dry. Snow seldom comes to the high forests clinging to the slopes, but when it does, the blizzards are fierce, and huge drifts build up within hours. This is the kingdom of the 'snow monkey', a name given by the local Lahu people to the Yunnan snub-nosed monkey. The monkeys range from 3000 to 4500 metres (9840–14,765 feet). At this altitude, the air is very low in oxygen, and so humans paying fleeting visits run a high risk of altitude sickness. This is the limit of where trees can survive, and good browse is scarce, yet the snub-noses can be seen in huge groups – sometimes almost 300 strong. The survival of these social monkeys depends on another high-altitude specialist, a species of lichen that hangs in beard-like drapes from the branches of larches above 4000 metres (13,125 feet) and makes up more than 70 per cent of their diet in winter, when the snub-noses' daily routine consists of little else other than searching for the lichen.

These highly specialized monkeys are confined solely to the highest slopes of these few mountains and one more over the border in Tibet – all part of the Bai Ma Xue mountain range in the middle of the Hengduan mountain region. Life at such high altitudes can be incredibly harsh, especially in winter, when night temperatures can plummet well below -30°C (-22°F). That the monkeys are stranded here is a product of geological events that have defined the region.

Tumbling down from the Himalayan plateau, three of Asia's greatest rivers – the Yangtze, the Mekong and the Nu Jiang (Salween) – cut north to south through

northern Yunnan. Their gorges – three of the deepest in the world – run parallel, just 30km (19 miles) apart, and seen by satellite, they appear clawed into the land. The courses of these rivers were more or less set in place some 50 million years ago, by the same processes that caused the Bai Ma Xue Shan to rise.

India was once an island, the high part of a tectonic plate moving slowly northwards on a collision course with Asia. As the two landmasses met, seafloor sediments began to rise, and to the north, the Himalayas lifted nearly 10km (6 miles) high. To the side of the collision, Yunnan buckled and folded, and row upon row of long, straight, immense mountain ranges rose up parallel. US geographer Pete Winn suggests pushing your hand across a tablecloth to get an idea of the effect: the wrinkles in front of your hand are the mountains of Tibet, and the creases to the right are the Hengduan ranges.

The event lasted millions of years and created dramatic changes. Yunnan was carved into slivers of isolated land (Hengduan means 'cut across' in Chinese), separating populations of animals and plants. Both the frozen mountain-tops and the treacherous gorges and rivers below acted as barriers. As the Bai Ma Xue Shan ridge slowly rose between the Mekong River and the Nu Jiang (Salween River), the ancestors of the snow monkeys were cut off from other snub-nosed monkey populations. Over time, they specialized to best exploit their territory, eventually becoming a distinct species.

BELOW
The famous Himalayan blue poppy *Meconopsis betonicifolia*. It was first noted growing wild in Yunnan by Catholic missionary Père Delavay in 1886 and found fame in England when seeds brought back by the botanist-explorer Frank Kingdon-Ward were germinated, becoming the horticultural hit of 1926.

plant-hunters' paradise

Just a slight change in altitude sees a dramatic change in the species of both trees and the plants in the understorey, and so the incredible altitudinal and climatic range of the precipitous slopes of the Hengduans has created endless opportunities for diversification. The region is a botanical paradise. Throughout spring and summer, the slopes of the Bai Ma ridge are a riot of colour as a succession of plants come into flower. In the forests, banks of different rhododendrons bloom flamboyantly in bright pinks, reds, yellows and fluorescent whites. The foliage is just as diverse, from delicate silvery leaves covered in fine down to brash, intensely green broad leaves that seem almost tropical. At high altitudes on the slopes of Dali Chan Shan, at the southern end of the Hengduan Shan, the flowering of a forest of alpine azaleas and camellias creates a spectacle famous throughout China.

Endemism is incredibly high. Approximately 700 flowering plant species including primulas, gentians, anemones, clematis, magnolias and lilies are exclusive to the area – lots of them familiar to us as garden flowers. In fact, many of the classic western garden favourites have their roots in Yunnan.

In the 1930s, James Hilton's novel *Lost Horizon* gave us one of the great myths of modern times – Shangri-La, an isolated mountain utopia where time stood still. His inspiration came from the exploits in the early 1900s of the great plant hunters of Yunnan. These eccentric botanical explorers came to the valleys of Yunnan in search of adventure and new species. This part of China was regarded as a lawless territory peopled by fierce tribes, some of whom, like the Wa people, even practised headhunting. Travelling over incredibly difficult terrain, the plant hunters entered valleys that few outsiders had, enduring many adversities, encountering new cultures and getting into scrapes with bandits. Their reward was the discovery of one of the richest temperate floras in the world.

The adventures of Joseph Rock, an Indiana Jones of the botanical world, are the most well documented. Rock was a fiery loner, who sent tales of daring-do and new frontiers back to the National Geographic Society. In his book *The Great River Trenches of Asia*, published in 1925, he described fantastic journeys crossing hair-raisingly high passes, encountering blizzards one day and leech-infested forests the next. He became the first westerner to cross the now world-famous Tiger Leaping

Gorge, describing it as the finest of all the gorges in Yunnan. Climbing up to the ridge looking into Burma, he met Lisu warriors, their arrows tipped with poison. By night, he reported, they used bamboo fire crackers to keep tigers away from camp.

Just like the characters in Hilton's novel, Rock fell in love with Yunnan. He was fascinated by the tribes he encountered, becoming particularly dedicated to the local Naxi people of Lijiang, close to the Tiger Leaping Gorge. He settled there for almost 30 years, studying and documenting the Naxi's animist, matriarchal culture. Becoming fluent in their pictographic language, and with the help of the shamanistic priests, or dongbas, he meticulously translated as many of their historical records as he could, convinced that the Naxi culture might not endure as Yunnan finally began to open up to the turbulent China of the 1930s and 1940s.

The other major legacy of the plant-hunters' expeditions was the discovery of an immense abundance of new species of trees, herbs and other flowering plants on northern Yunnan's slopes. The collections they made and shipped home were astounding. Rock sent thousands of species to arboretums in the US. George Forrest, a Scottish botanist who explored the area in the early 1900s, sent more than 31,000 herbarium specimens to the Royal Botanic Garden, Edinburgh. On one collecting trip in Yunnan, he narrowly escaped death at the hands of Tibetan warrior priests, who attacked and killed the rest of his party. He escaped, tumbling down into a valley, and was hidden by Lisu people, continuing later with his collecting.

ABOVE
A young Tibetan hunter photographed by Joseph Rock on the banks of the Mekong River in 1923 – one of hundreds of pictures he took of the tribal people of Yunnan.

He is credited with the discovery of more than 50 species of primulas, lilies, gentians, camellias, clematis, irises, jasmines and many more now common garden species. It is for the rhododendrons, though, that he is most remembered, having collected 300 or so species, including *Rhododendron forrestii*, a ground-hugging species with huge red flowers and *R. clementinae*, a species he named to commemorate his wife.

Yunnan is an epicentre of rhododendron and azalea evolution. Some 465 species are native to the area, ranging from immense 30-metre-high (98-foot) rhododendron trees, found only in the most remote forests, to the ground-hugging alpine mats of jewel-flowered azaleas that can be seen in garden rockeries throughout the west. The moist temperate climate of the Yunnan highlands is not, in fact, dissimilar to that of Britain, and the discoveries of the plant hunters such as Joseph Rock, George Forrest, Frank Kingdon-Ward and Ernest Wilson bloom annually in gardens around the world.

forest of riches

Joseph Rock's journey, travelling west across the Yangtze valley into the Mekong and finally over into the Nu Jiang, would have allowed him to experience the full range of Yunnan's climates. Descending into the Nu valley in October 1924, he noted the dramatic contrast between the steamy heat there and the cool, drier highlands of the Bai Ma Shan ridge he had left only a few days before.

Of the three great river valleys, Nu Jiang is by far the lushest. Its deep trench runs below the imposing Gaoligong Shan, which forms the border with Burma. 'Nu' means angry in Chinese, and even in the dry season, the Nu Jiang waters rage. On either bank, tiny hamlets of stilted houses of the Nu and Lisu people sit precariously on the steep slopes among a patchwork of terraced rice and sweetcorn fields. Impossibly thin cables stretched across the valley link the hamlets. On market days, there can be queues of people waiting to whirr across the river at speed.

Joseph Rock used the same method in 1925 to transport his entire entourage, including 14 mules from bank to bank. The cables then were made from twisted bamboo: 'Both rope and slider are greased with yak butter to facilitate the crossing,' he noted. But it wasn't failsafe, and two of the mules were swept away by the surging waters. Multitudes of springs feed the Nu, tumbling down from the high canyons of the Gaoligong Shan. These valleys are some of the least explored parts of China, often

BELOW
A Naxi man and woman in everyday clothes, the man smoking a pipe made from bamboo. The Naxi people are believed to have migrated from Tibet in the first century. Traditional Naxi houses are two-storeyed wooden structures, often with carvings on the doors and windows.

impossibly thin cables stretched across the valley link the hamlets.
On market days, there can be queues of people waiting to
whirr across the river at speed

lost in swirling clouds and cloaked in the densest, richest forests of Yunnan. Mosses drape from the twisted boughs of giant rhododendron trees found nowhere else. Tropical climbers twist around tree trunks, whose branches seem impossibly laden with the most incredible collection of orchids, their pseudo-stems like thick strings of green beads entangled in the moss.

Jewel-coloured sunbirds dazzle in the eerie half-light as they dart through the bamboo undergrowth and up into the canopy. Here they flit between blousy, scarlet blooms of giant rhododendrons and tiny pink flowers of hanging orchids – epiphytes that live on the branches, obtaining moisture from the air. Constantly on the move, they are easy to pinpoint by the sharp trill they make as they dash from flower to flower sipping nectar. Often they are accompanied by flocks of as many as a dozen species of other tiny forest birds.

The forests support the highest density of primate species in the northern hemisphere. Phayre's, or spectacled, leaf monkeys are the quietest and perhaps the most beautiful, their faces a delicately coloured mask. They work their way through the canopy, avoiding the poisonous rhododendron foliage and plucking young broadleaf shoots. The duets of family groups of hoolock gibbons – the most northwestern of all gibbons – can be heard each morning. The undergrowth belongs to troops of macaque – three species, the largest and most impressive being the red-faced, or bear, macaque, known by the local Yi and Bai people as big-foot monkey.

The forests of Gaoligong Shan are home to most of Yunnan's pheasant species, including Lady Amherst's and flocks of white-eared pheasants – more commonly thought of as alpine-forest animals. The thickets of bamboo are the stronghold of the much rarer Temminck's tragopan. The tragopans are incredibly shy, and even the male's mating display is limited to a brief but startling game of 'peek-a-boo'. Instead of displaying with flamboyant feathers, the male stalks his female until he finds himself close to a log or hump. Ducking behind the cover, he starts making a sharp clacking noise. As he clacks and wing-flaps himself into the mood, a lappet – a long

ABOVE
A Lisu woman crosses the Nu in the traditional way. Not long ago, the only roads were pony tracks. The Three Parallel Rivers of Yunnan Protected Areas, of which this is one, are swathed by relatively untouched forest, with a diversity of life, including more than 25 per cent of the world's animal species.

flap of coloured skin – falls from below his chin and begins to stiffen, as do tiny blue horns on his head. The clacking gets quicker and louder, and the horns get stiffer and longer until, in a flutter of wings, he rises up to more than twice his normal height, presenting the startled female with his fantastic bib of neon blue and red.

On the slopes of the Gaoligong Shan is a fantastic mix of subtropical, temperate and alpine plants and animals, all thriving in the moist, mild climate. Deciduous oaks, maples, wild cherries and hemlocks grow alongside fig trees, palms and tree ferns. The conditions that allow for this transition through biogeographical zones

LEFT
The dramatic finale to the mating display of the male Temminck's tragopan. His striking bib, or 'lappet', is shown in its full glory for just seconds as he plays peek-a-boo with the female.

are unparalleled and have resulted in 10 per cent of the world's flora being found here and only here. With each plant-hunting expedition, new species are found. Recently, teams from the US and China together collected more than 11,000 specimens, many of which may add species to the 4303 seed plants already recorded from here. Their entomologist colleagues counted around 1690 species of insects.

OPPOSITE
Mosses and lichens drape the trees of a species-rich mixed forest lining one of the many moist, warm, deep river valleys in the mountains of southwest China. The region is a biological hotspot of great conservation value.

BELOW
Male golden pheasants on their lek (dance ground), displaying their capes of feathers and flamboyant tails to a female.

warm winds

There is a mystery to this abundance. At this latitude and in such a land-locked area, the climate should be much colder. It should also be a good deal drier, chilled by winds blowing down from central and northern China. But as the leeches verify, there is no doubt this is true rainforest. So how does such an incredible plant paradise exist so close to frozen Tibet, so far from the tropic of Cancer?

The key is the way the mountains and valleys all run north to south. The two huge mountain ranges of Ailao Shan and Wuliang Shan shield Yunnan from the cold, dry northern continental winds. Below them, the narrower river gorges of the Mekong and Nu Jiang run south and open wide to the humid tropics of Southeast Asia. When the annual monsoon rolls in from the Indian Ocean, the valleys act like enormous funnels, channelling the warm, moisture-laden winds northward.

In about the last week of March, gales reach the forests of Gaoligong Shan. They shred the leaves of wild bananas, send orchid-laden branches crashing to the ground and bring down showers of scarlet rhododendron flowers. Sunbirds and their fellow nectar-drinkers work quickly in the morning when the wind is still a light breeze. By the afternoon, they must snatch feeds from flowers in the understorey, as higher branches toss and sway in powerful gusts of warm air that signal the arrival of the monsoon.

Through May, June and July, rain drenches the high mountain slopes and their rich valleys. The immense rivers swell and rage southwards, and the Nu is then said to be at its angriest. But most of the water is kept in the hills by the forests, and the

PLANTS FOR ALL PURPOSES

The most important plants in Chinese culture are the 500 or so species of bamboo, 400 or so of which are found in Yunnan. Bamboo forests have inspired artists through the ages. But bamboo is also a fundamental part of daily life, playing a significant economic role.

Bamboo is so supple that it can be woven into cloth, so fleshy that it makes delicious dishes, so strong that it is used in scaffolds to support Hong Kong's high-rise towers (it is stronger in tensile strength than steel). It is the ultimate utilitarian plant. Baskets and mats have been made from it for at least 5000 years, bamboo paper was invented about 1100 years ago, and long before that, characters were scratched on slips of green bamboo – possibly the first writing. Wars were fought with bamboo bows and arrows, and with the Chinese invention of gunpowder, bamboo firearms and missiles were used. Today, bamboo fireworks cheer in the Dai new year.

Bamboo originated in the tropics but has migrated into even alpine zones. These giant grasses are the fastest growing woody plants in the world – some species have been recorded as growing more than a metre a day – and a single clump can produce up to 15km (9.3 miles) of usable pole in its life.

Though the shoots are soft, mature bamboo is high in indigestible fibre and loaded with abrasive silica, making it difficult to eat and to digest. Some believe that the hard stems and leathery leaves evolved to deter grazing by dinosaurs (bamboo fibres have been found in dinosaur droppings).

A few animals specialize in eating bamboo, most of which are found in China. The giant panda is the most famous, but its smaller cousin the red panda has a similar diet. They share the same habitat by feeding in slightly different ways – the red panda climbs up trees and along branches to reach the softer green bamboo leaves, while the giant panda crunches through the tougher bamboo culms.

Attack comes from below, too. Stand in a forest for a time, and you will soon see the odd bamboo stem shaking and, bit by bit, disappearing below ground – into the burrows of bamboo rats (left), who spend their lives underground sniffing out the bamboo growing in the forests above.

these giant grasses are the fastest growing woody plants in the world – some species have been recorded as growing more than a metre a day

deep valleys sustain a warm, moist atmosphere throughout the year, releasing the water in gentle streams that feed the lowlands. Evaporating water forms fog and mist each night, condensing in the morning as dew and rain. A popular saying is that they have 'a dry season without the drought'. Even in periods of global change, these forests have maintained an environment of great stability.

f o r e s t   r e f u g e s

Millions of years ago, the world went through a series of dramatic ice ages. Immense ice sheets scoured the land, and during the hotter years in between, sea levels rose dramatically, flooding huge areas for thousands of years. Sitting so close to the equator and being so influenced by the moist monsoon, the climate of Yunnan, along with that of much of Southeast Asia, remained comparatively unaffected. While many of the world's rainforests disappeared, rainforests in this corner of the globe survived – becoming for many species the last sanctuary. Frozen out elsewhere, most of the great apes survived in the protected valleys, and it is thought that the forests of Yunnan may have been one of the last refuges for the ancestors of orang-utans and gibbons. As the planet came out of the ice ages and warmed up, the forests acted as reservoirs of wildlife from which species could spread again.

BELOW
A red panda feeding on bamboo, which constitutes more than 90 per cent of its diet. Known in China as the fire fox, the red panda is still found in Sichuan and Yunnan, but over the past 50 years, its numbers have plummeted by around 40 per cent.

Gaoligong Shan's year-round mild climate also played an important role in one of the earliest world trades. Practically the only way to access the forest here is by cobbled pathways that wind up and down through the many valleys to the borders of Burma and beyond. In places, the cobbles are interrupted by boulders with tiny, polished steps worn deep into the rock. The little paths are part of the centuries-old southwest tea and silk route along which traders from as far as Rome brought their wares. About the same time that silk appeared in Rome, Buddhism arrived in China. Today, the immense valleys remain important corridors linking the north to the south. Thousands of birds, including black-necked cranes and Derbyan parakeets, migrate along Yunnan's valleys – a direct route from northern territories in Tibet down into China's last remaining tropical rainforests. The valleys of southern Yunnan's Xishuangbanna, with its unique mix of temperate, tropical and subtropical species, are a hotspot of diversity. They are at the northernmost edge of the monsoon zone, but seasonality affects them in different ways. Each valley has its own cycle of mini-wet and mini-dry seasons, and the forests differ from one valley to the next. In one, dipterocarps – immense tropical trees – tower above the canopy signifying true rainforest. In the canopy layer, climbing bamboo and twisted lianas spiral down from high branches, while palms and rattans contribute to the tangle of understorey below. In the next valley, tree ferns or Nepal cycads might dominate the steeper, exposed slopes creating a primeval scene. Further from the tropical border, towards the Nu Jiang and Mekong valleys, more and more temperate species begin to appear. In the mini-dry seasons, bare branches can be seen above the canopy where temperate trees push through alongside the tropical evergreens.

Not only are the forests here on the fringes of two ecological zones, but they also stand on the junction of two immense land plates. This is reflected in the diversity of plants. Thousands of new species arrived with the Indian plate as it made contact, and it is thought by some that bamboo may have first arrived in China this way. Bamboos are incredibly diverse in Yunnan – more than 400 species have been counted – though scientists report that local people make use of many more species that have not yet been given scientific names.

loud, smelly nights

Though the forests harbour an abundance of wildlife, it can often be incredibly difficult to see – but you can hear it. Sound can be the best way for forest creatures to communicate, whether to protect a territory or to find a mate. But calling attracts predators as well as mates, which is why much of the action happens under the cover of darkness. Indeed the night can often be the loudest time in the forest.

The forests of Xishuangbanna are home to one of the world's few species of gliding frog. Reinwardt's tree frogs normally spend their time in the highest branches of the canopy hunting the abundant insect life, but in late April, they glide down at night to gather on branches overhanging small forest pools. The males compete in a night chorus, singing for the chance to mate – their songs joining those of the many other frog species doing likewise. Groups of up to 20 gather around the larger females, and a mating scrum often ensues as the males try to dislodge competitors. Once a suitor is firmly in place, the female builds up a huge

BELOW
Yunnan's remarkable Reinwardt's tree frog. The skin between its large toes allows it to steer its descent as it glides between branches in the forest canopy.

foam ball on the tips of leaves overhanging the pool. After two weeks inside the foamy nest, the tadpoles wriggle their way free and drop into the pool below.

It is at this time of year that the rainforests of Yunnan become particularly smelly. Wind doesn't penetrate far into the tangles of the rainforest. So plants here tend to rely on jungle animals for pollination, using scent as an attractant. Gloriously heady aromas of exotic flowers fill the night air, attracting moths to pollinate them with the promise of sweet nectar. But not all such scents are as pleasing to human senses. Underground at the edge of clearings, great tubers, often up to 10kg (22 pounds), lie dormant. Come the rainy season, a huge, grotesque 'flower' erupts from the ground. Called 'the witch of the forest' by local Hani people, the 'corpse flower' of the elephant-foot yam transmits a concoction of repugnant smells.

As the sun descends, the purple warty 'petals' (bracts) peel away to reveal a huge central column – the spadix – which inspired the plant's group scientific name *Amorphophallus*, meaning shapeless penis. From it the most incredibly putrid stench begins to emanate, permeating far and wide with incredible speed. If you could bare to get close enough, you would notice that the plant is positively throbbing with heat. In fact, the temperature of the spadix rises by an incredible 10°C (50°F). The heat is thought to be the by-product of the chemical reactions taking place inside

ABOVE
The corpse flower of the elephant-foot yam – a source of disgusting smells and valuable food. The yam is thought to have been brought to Yunnan as a forest crop many thousands of years ago.

the spadix to create the stink, but it may also help the scent to carry further and faster. The combined effect of the smell of rotting flesh, the heat and the purple fleshy 'petals' makes for a convincing illusion of death. The carrion beetles arrive in droves from the surrounding forest and literally fall into the plant's trap.

As the beetles land, they tumble down the waxy slopes of the petals into the heart of the structure and cannot escape. At the base of the spadix are tiny true flowers waiting to be pollinated. With luck, some of the captive beetles will have visited other corpse flowers beforehand and will be carrying their sticky yellow pollen. To allow time for the pollen to rub off onto the flowers and pollinate them, the beetles are held for 24 hours. As a reward for captivity, some of the minute flowers provide protein parcels – a beetle-sized packed lunch. Then, as the sun sets again, the flowers release their own pollen, covering the trapped beetles. Finally, the fast-decomposing petals fall open and the insects can fly free to carry the pollen to more big smelly flowers.

food from the forest

The elephant-foot yam is widespread through Yunnan's forests. It grows in forest clearings and is often found in small groves close to hill-tribe villages. In fact, moyu,

BELOW
A typical array of fresh vegetables in a Yunnan market. Often the spices, herbs, fruit and vegetables originate from the forests of Yunnan or the forest gardens that are still cultivated by many villagers.

THE TREE-CLIMBING FISH

During the research for the *Wild China* television series, a remarkable photograph was found in a book in a hotel near the forests of Gaoligong Shan. It seemed to show two catfish-like animals clinging to the high branch of a tree. Eventually, the village where the picture was taken was located – and so was a unique behaviour. The locals told how, in May, thousands of large fish could be seen engaging in a mating frenzy in the trees overhanging streams near the village. Could they be laying their eggs on the branches? The bad news was that a dam had just been completed on the one river where the fish were known to live.

A two-day journey to the river took the researchers close to the border with Burma, where an interview with a local Lisu tribesman revealed the truth. The fish was indeed to be found there and thousands did emerge at the onset of the wet season. Though they don't lay their eggs in the trees, they do climb over boulders and trees as they move upstream along the streams that feed the river to reach the higher reaches, where they lay their eggs. Though proud of their unique fish, the locals still catch them – they are said to be very tasty   but strict traditional rules forbid disturbing the fish on the journey to their spawning ground. Only once they have laid their eggs and are returning to the river can they be caught with bamboo fish traps.

The other half of the story was the dam. Completed in the winter of 2006, it provides a regular supply of electricity to the village. But the following spring, for the first time in history, the tree-climbing fish spectacle did not occur. The water level of the river is so low that the fish would have to cross too great a distance out of water to reach the higher levels of the streams, and it is highly likely that this remarkable fish – still without a scientific name – will soon exist only as a legend.

in May, thousands of large fish could be seen engaging in a mating frenzy in the trees overhanging streams near the village

as the yam is called locally, probably came from Malaysia but was spread throughout Asia thousands of years ago as a forest crop. It is actively cultivated by many of the minority people of Yunnan. The huge tubers are dried and crushed to a powder, which when reconstituted with water, sets into a firm, clear jelly. Stir-fried with the local mouth-numbing 'gunpowder' spice, it makes a delicious and unforgettable dish. The elephant-foot yam gel has also been discovered by the international food industry and is now used in all sorts of products, from tortillas to ice-cream.

BELOW
A Hani child with the family's water buffalo, used to work their rice terraces. The Hani traditionally live in the mountain areas around the Mekong and Red rivers and are just one of 25 ethnic minority groups found in Yunnan.

the Hani knew ... that the forests gave rise to the streams that watered the lowlands ... Remove the forests, and the crops would dry up

Many villages grow crops of wild vegetables within the forests, some of which would be familiar in our kitchens. Yunnan's forests are an epicentre of ginger diversity, with more than 200 species. A closely related cardamom species also does well in the shady environment of the forest. Many species of wild bananas originate from here, including one giant that is 4 metres (13 feet) tall and has flower petals more than 80cm (31 inches) long.

Minority groups such as the Jinuo, Bai and Yi were traditionally never farmers but rather nomads within the forest, practising slash-and-burn cultivation alongside hunting. They followed a pattern, set by tribal elders, of settling in the forest for periods of about ten years. They would then move on, leaving the ground to be reclaimed by the forest.

Slash-and-burn agriculture has been regarded as incredibly destructive, but some studies have shown that, on a small scale, 'swidden' agriculture can actually increase the diversity within a forest by opening up small patches for 'pioneer' plants – the first to recolonize a disturbed area.

Bananas, gingers, bamboos and elephant yams are all pioneer plants. Fast growing, often with soft, fleshy vegetation or nectar-rich flowers, they can support a large number of insects and birds and are the preferred browse of many of Yunnan's forest animals, from the tiniest mouse deer (the world'smallest deer) to the largest forest dweller, the elephant.

In 1999, a national logging ban came into force in an attempt to halt deforestation, which the government rightly linked with massive erosion and flooding. Though this has protected some forests from widespread destruction, illegal deforestation by logging companies still occurs. Meanwhile, the ban has also meant the end of a way of life for many of the hill peoples of Yunnan and an irreversible change to their cultures.

the golden kingdom

By the time the Mekong River reaches Yunnan's tropical south, it is swollen and meanders lazily through wide valleys. Here and there giant bamboos lean in towards the slow-moving water, but the only true natural forest remaining is on the rolling hills and mountains above. Along the banks of the river, the rich and fertile lowlands have long been dominated by the rice paddies and vegetable plots of Yunnan's master horticulturalists, the Dai people.

In the ninth century, the Chinese adventurer Fan Chuo brought news of his escape from Vietnam through an exotic 'golden kingdom' on China's southern borderland. He told of immense rice terraces where oxen and elephants were used to till the land. Its people wove intricate materials, made weapons of metal and even plated their teeth with silver and gold. This was the kingdom of the Dai.

The Dai are valley people. Verdant rice fields surround the streams and rivers that flow through their villages, and every household tends vegetable gardens packed with strange tubers and gourds. Built on some of the most fertile lands in China, the Kingdom of the Golden Temple was reigned over by hereditary kings for more than a thousand years and was the northernmost of three Thai kingdoms (the others would later become Laos, Thailand and Burma).

Chinese emperors longed to gain control of Yunnan. The abundant lowlands could feed their armies, but more crucially,

ABOVE
A typical Dai house, with a maize storage area and a vegetable garden in the front.

they needed access to that high, winding path through Gaoligong Shan forest. This was the gateway to the world beyond the Middle Kingdom, with its hunger for China's silk and tea and all its knowledge about the Buddhist faith. This tiny forest road could bring exotic riches, wealth and new power to the empire. In 109 BC, Emperor Wu sent armies into Yunnan to re-establish the western silk and tea route, naming it the South West Barbarian Way. Over the next thousand years, repeated incursions were made in attempts to maintain power, the most successful being that by Kublai Khan in the thirteenth century. With each attack, waves of Dai immigrants would move south into what is now Thailand and governors would be appointed to administer the territories. But they were never wholly successful. The sheer distance – more than 3220km (2000 miles) from China's capital – the strength of the Dai culture and the impossible Yunnan terrain all served to ensure that the Dai social hierarchy, with its local princes and land taxes, continued much as it always had. Not until 1953 was the last hereditary king of the Dai finally removed from power. He was made an official by Mao's army and, to this day, lives in Kunming, Yunnan's capital.

Since then, Xishuangbanna has changed phenomenally, but the Dai ethnic group is still the largest and wealthiest in Yunnan, and in many towns, it would seem odd to call them a minority. Even in the biggest Dai city, Jinghong, Dai women can be seen in their distinctive shimmering wrap dresses. They wear their hair high in tight knots, often with wraps of startlingly bright silk. Heavy, elaborately patterned gold studs decorate their ears, and gold-capped teeth are as common in the suburban and rural villages now as they were when Fan Chuo passed through. The rich golds of their pagodas are distinctively Thai. Dai traditions and festivals would be very familiar to a Thai family, and the water-splashing new year festival is the biggest celebration of the year both in Thailand and Yunnan.

ABOVE
A Dai woman selling parcels of spicy food wrapped in leaves. The food in this region is unforgettable.

The most startling reminder of how close these cultures are is the unforgettable food. Fish wrapped in leaves are steamed with complex concoctions of wild herbs and home-grown spices. Meats are marinated with mouth-watering results and served in delicately flavoured wild-banana-flower soup. Even a bowl of spiced ants mixed with fresh-chopped forest herbs is delicious.

Though they tend the most impressive vegetable plots and are famous for their rice, much of the flavour of their cooking comes from the forests on the hills above. Dai villagers still know what the forests offer, and Dai women regularly collect sharp-flavoured herbs to sell in the local markets and use at home. The fact that forests still remain may well be due to ancient traditions. Sacred forests above settlements were protected by strict systems. They were often subdivided – into cemetery forests for burying the dead, forests for cutting wood, forests for gathering herbs or bamboos – and overseen by a village elder, who could deal out punishment to those who flouted the law. Other minority groups, the Hani, for instance, protected their forests in similar ways, knowing they were a vital component of their culture and welfare. They also knew that the forests gave rise to the streams that watered the lowlands throughout the year. Remove the forests, and the crops would dry up.

empty forests

'If we destroy the forest, it will become desert. Our Communist Party of China will become criminals in history, and future generations will curse us,' Premier Zhou Enlai is reported to have said on a visit to Xishuangbanna in 1961. Now new forests have come to dominate many of Yunnan's southern slopes. Silent and regimented, they are made up of row upon row of Brazilian rubber trees. These, far more than the swidden agriculture of the Jinuo and Hani people, have been the biggest threat to Yunnan's incredibly diverse and unique forests.

Native to the Amazon, rubber trees produce sticky sap, latex, which is processed and moulded into the tyres that make the wheels of the cars of the world go round. The industry is of immense value, and in the 1950s, Chairman Mao was keen to exploit this, ordering vast tracts to be planted. But while rubber plantations have helped the growth of the Chinese economy, it has been at a massive cost to the environment. In a native forest, where the moist climate changes little from day to day, there is immense plant and animal diversity. In a rubber plantation, there is only one species, the Pará rubber tree. The land beneath rubber trees is kept clear of other plants, scorched by sun and then exposed to pounding heavy rains. There is no rich layer of leaf-litter to feed the plants or absorb rainwater. Heavy fertilizers are needed to maintain high yields, but under monsoon rains, the fertilizers, along with the bare soil, are soon washed away. The plantation canopy, devoid of layers and lacking the rich mix of lianas, mosses, orchids and ferns, lets any evaporated water be lost to the atmosphere. This dramatically changes the climate on a large scale. Around the Jinuo minority valley close to Jinghong, more than 40 streams dried up as a result of clearances for rubber. The area was also subject to greater temperature extremes, less rainfall, stronger winds and a much drier climate.

BELOW
Devastation caused by logging native tree species on the Bai Ma Xue Shan. Despite massive losses, Yunnan still has some of the biologically richest forests in China.

Rubber plantations continue to spread in Yunnan but have been overtaken by yet another crop. Eucalyptus is being planted throughout southeast Yunnan, and the amount of eucalyptus planting in China is now second only to Brazil. There are also recent reports of natural forests being illegally felled to make way for these fast-growing aliens. Eucalyptus forests provide profitable timber but, like rubber tree plantations, make poor replacements for the riches of Yunnan's own unique forests.

local heroes

In the early 1990s, Yunnan snub-nosed monkeys numbered only 1000–1500. Heavy poaching was taking its toll, but the main threat came from the illegal logging that was occurring at a frightening rate on the Bai Ma Xue Shan, even though the area was officially a nature reserve. The last remaining habitat of the endangered snub-nosed monkey was about to be lost.

The monkey would surely have become extinct in the wild were it not for the actions of wildlife photographer Xi Zhinong (whose pictures are in this book) and his journalist wife Shi Lihong. Xi was spending weeks in the frozen Bai Ma Xue peaks, making a film about the monkeys, when he discovered that a logging company had been given local permissions to head straight into snub-nosed monkey territory. Zhinong took the extremely brave step of writing to the highest authorities. A copy of his letter was leaked to the press, and the story went international, aided by his photographs. The plight of the monkeys stirred the emotions of the public and inspired Beijing students and other environmentalists, among them Shi Lihong, to organize a 'green camp'. This involved a long trek towards the Bai Ma Snow Mountain Nature Reserve, holding meetings with government officials and villagers along the way. The pilgrimage caused a media frenzy and public concern. Zhinong's award-winning film about the monkeys brought a final end to the logging company's plans, and the government responded by ending the illegal deforestation. Since then, Yunnan snub-nosed monkey numbers have increased and have been the subject of studies. The story is now legendary, showing how, by using film, photography and the media, it was possible to turn a little-known animal into a national cause and by doing so, save both a species and a forest.

Currently, a little over 18 per cent of China is forested – a recovery from 1947, when just 8.7 per cent of forest remained. The reversal is due to a massive reforestation of 20 million hectares (49 million acres) between 2000 and 2005. But more than a third of this area is plantations. The government has recognized the need to protect the remaining natural forests and is now enforcing a national ban on logging. In Yunnan, more than 190 reserves have been created (more than in any other province). Recognized internationally as places of extraordinary biodiversity, these forests have once again become vital yet fragile havens to the incredible wildlife they harbour.

OPPOSITE
A family of Yunnan snub-nosed monkeys in relict larch forest high up in the Bai Ma Xue Shan. In winter, up to 70 per cent of their diet is lichen, which hangs from the branches of the trees.

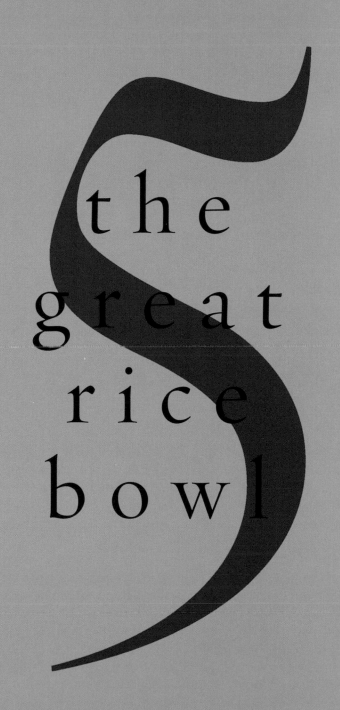

the
great
rice
bowl

IN YUNNAN PROVINCE, JUST NORTH OF KUNMING, the mighty Yangtze River – known up to this point as the Yalong Jiang, or Dangerous Dragon River – finally escapes from the mountains. It makes an abrupt hairpin bend northwards before arcing east in a massive 2500km (1554-mile) curve to reach the East China Sea at Shanghai. Below this vast river arc, southern China basks in a warm monsoon climate, which cloaks the seemingly endless ranks of hills in a near-permanent haze, producing the characteristic 'watercolour light' captured in so many classical Chinese paintings. About a fifth of the land is intensively cultivated, supporting an estimated 300 million rural inhabitants, most of them rice farmers. Hills and mountains comprise the remaining four fifths, offering some of China's most beautiful and memorable landscapes – none more so than the bizarrely sculpted karst hills of the Li River valley south of Guilin in Guangxi province.

To travel down the Li River on a misty morning is one of the most magical experiences you can have in China. The river slips between the cliffs of towering, teeth-like hills, and overhead wheel screaming bands of juvenile house swifts, while cauliflower clouds gather on the horizon in preparation for an afternoon downpour. At Xingping village, cormorant fishermen perched precariously on bamboo rafts dance and chant for the benefit of the tourists' cameras. The river here is too murky for the cormorants to see any fish that may be lurking among the rocks, but the fishermen have a supply in the baskets on their rafts, which they toss one by one into the water to display their birds' fishing skills. Even though it's all just for show, you know this vignette of traditional China will stay with you for ever.

the sculptured karst

The fantastical karst landscape covers more than a million square kilometres of southern China. Karst is a European term that refers to a landscape created by the action of water on limestone, and China has more exposed limestone than any other country – the vast majority in the south. The reason that karst hills assume such strange shapes in China has to do with the alchemy created when warm, subtropical rain falls on thickly bedded limestone.

As a raindrop forms and falls, it absorbs gases from the air, including traces of carbon dioxide, producing a solution of weak carbonic acid. In a subtropical climate, the activities of soil bacteria are speeded up, and so the rainwater gains even more acidity as it percolates through the soil. Weaknesses in the limestone bedrock, such as the vertical stress cracks produced by mountain-building movements in the Earth's crust, are transformed by the water into fissures, potholes and caves. Eventually, the water forms underground streams that drain the landscape internally. Meanwhile larger rivers flowing between the karst hills eat away at their bases, undercutting their slopes and so producing steep-sided cliffs.

PREVIOUS PAGES
Ripening rice on terraces built hundreds of years ago on the hillsides of Longsheng in Guangxi province.

OPPOSITE
Li River fishermen with their trained cormorants, framed by the famous limestone peaks of Guilin. Tourism provides the means to continue this traditional way of life.

FISHING WITH CORMORANTS

the oldest Huang is the trainer and keeper of 12 cormorants, whose eggs he incubates under a broody hen, hand-rearing the squabs so that they imprint on him

Steering three rafts of lashed-together bamboo stems using bamboo poles are Huang, Huang and Huang. Like most cormorant fishermen on the Li River, they are related and share the same surname. Dressed in working pyjamas and bamboo hats, each has four or five tethered cormorants on his raft. The oldest Huang is the trainer and keeper of 12 cormorants, whose eggs he incubates under a broody hen, hand-rearing the squabs so they imprint on him. The second Huang sets up photo-assignments with tourist operators. 'Young' Huang has never learned to read or write, and so has little choice of other work locally.

Cormorant fishing is an ancient art, passed down from father to son across most of the major river systems of southern China. A cormorant is a master fisher, grabbing prey in its long, hook-tipped beak, then bobbing up to the surface to gulp it down. All a human needs to do is fit it with a collar to stop the fish going down and train it to return to the raft.

As dawn breaks over the jagged skyline, the Huangs steer their rafts into a shallow oxbow. There is a flurry of noisy protest as string is tied round the necks of the cormorants. Then the Huangs untie the birds' legs and begin to dance – knees bent, chanting and splashing. This is a signal to the birds to plunge into the river and search for fish. The activity intensifies as one surfaces gripping a flapping fish by the head. The birds begin to work as a gang. A cormorant appears with a fish, hotly pursued by three more trying to steal its catch. Young Huang extends his pole and calls to his bird, swings it on board, grabs it by the neck and dangles it over his fish basket until it regurgitates its prize. After an hour or so, the cormorants seem to lose motivation. One by one, they clamber back onto the rafts.

But these days, most fishermen rely on nets. The cormorants' real value is as a tourist attraction – the Huangs are professional entertainers. So what's in it for the birds? Trainer Huang demonstrates, rewarding his star cormorant with a catfish, removing its collar so it can swallow the fish. 'Like us,' he says, 'they do their work, but must wait for their pay until the end of the day.'

cavern life

For those brave enough to follow the internal plumbing of these hills, the rewards are great. China has more natural subterranean caverns than anywhere else, several thousand kilometres of which are of navigable dimensions. Some caves contain great roaring underground river rapids and waterfalls, others vast echoing chambers filled with tiered stalagmites up to 50 metres (164 feet) tall. Within these innumerable caverns live a vast array of cave animals found nowhere else – including several representatives of the carp family known as golden-line barbs. Some of these fish lack eyes and possess strange unicorn-like projections on their foreheads, the function of which (if any) is unknown. There are also cave-evolved crabs, shrimps, centipedes, millipedes and other invertebrates – many yet to be classified – and Chinese scientists are discovering new species each time a new cave system is explored.

Fortunately for those of us who have no wish to plunge into roaring underground rapids, the Chinese have developed a great many of their finest caves as

ABOVE
The karst scenery of Guilin, one of China's most popular scenic areas, also rich in plant and animal life.

tourist attractions. As a visitor to the famous Huanglong (Yellow Dragon) Cave near Zhangjiajie, in Hunan province, a concreted walkway leads to a jetty from where electrically powered flat-bottomed boats glide you along a clear subterranean lake into the heart of the mountain. At the far end, concrete stairs with handrails meander through a series of vast chambers hundreds of feet high and wide, decorated with icicle-like stalactites, glittering tiers of crystalline calcite terraces and shallow pools reflecting forests of slender stalagmites up to 5 metres (16 feet) tall – all lit with a galaxy of coloured fairy lights. The pleasantly mild air is filled with the excited chatter of fellow tourists – 99 per cent of them Chinese – and you are left with a spooky sense that Santa could well be lurking up a gloomy side-passage with a gaggle of Christmas elves.

By dusk the guides have flushed out the last of the tourists, the steel gates have been padlocked and the lights switched off. But the pitch-black Yellow Dragon Cave is far from deserted. The tunnels now echo to the patter of different feet, as white-bellied cave rats emerge from their hidey-holes to scavenge the crumbs left by the tourist hordes. The rodents find their way by following well-worn scent trails laid down by generations of rats and using their long whiskers as sensors. This is much the same system used by cave-dwelling invertebrates such as cockroaches and camel crickets, which use their extremely long antennae to feel out not only obstacles but also predators such as huntsman spiders, which lie in ambush in the darkness.

ABOVE
Furong Dong show cave in the scenic Three Gorges area of Wulong, Sichuan. Show caves may have loudspeakers and lights, but they are the easiest way to see the limestone riches of China's caves.

land of bats

A radically different remote-sensing system is used by another group of cave inhabitants. Bats produce ultrasonic squeaks in flight and use the pattern of reflected echoes to avoid obstacles. But echolocation is more than just a navigation aid.

On a nearby river, small fishes are patrolling just below the surface, on the lookout for drowning insects, their dorsal fins creating tiny ripples. In the dark, they are safe from prying eyes. But flying low to the water is a hunter with no need of light – Rickett's big-footed bat. It shouts out a barrage of ultrasound, decoding the echoes from the water surface and wheeling sharply round towards the source of the ripples. As it approaches its target, the bat lets down its undercarriage – outsized, forwards-facing feet, tipped with sharp, hooked claws – impales the fish and carries it away to an overhanging tree branch, where it munches it head-first.

Near Guiping in Guangxi province is a cave that echoes to the squeaks of tens of thousands of baby European free-tailed bats, tightly clustered in a wriggling tapestry of tiny pink bodies. Their mums are away – as far as 50km (30 miles) – in pursuit of migrating moths and beetles at heights of almost 3000 metres (9840 feet). A large bat colony can consume several tons of insect pests in a night – good news for local farmers. Small wonder then that bats are regarded as symbols of good luck. But there's a more prosaic reason: bat in Mandarin – *bian fu* – sounds like 'good fortune' or 'happiness'. So for centuries, bat symbols have been carved into jade and ivory and have adorned the façades of palaces and the thrones of emperors.

Come the dawn, each female bat will be back in the cave, searching out her baby in the squirming nursery. She'll track it down for its feed using her spatial memory of where she parked it, coupled with a nose for her offspring's personal smell and an ear for its unique squeak – all this amid the deafening chaos of 100,000 or more shouting mums and squeaking babies.

the troglodyte people

In Guizhou province, the morning marks an equally busy time for another cave dweller. Her mum is a long way away, and so she has to fend for herself. She has made a small fire and is busy boiling a kettle to make rice porridge. Her name is Liu and she is nine years old. There are other people in the cave – 18 families in all, with their cows, pigs, goats and chickens housed in neat wooden pens. Each family dwelling is neatly constructed with walls of woven bamboo. The houses lack roofs. There is no need to keep out the weather – a vast cave ceiling does that job.

Breakfast over, a dozen or so children play basketball on a purpose-made court. More children come running into the cave. Liu greets them as they remove their dripping pink, see-through plastic rain-capes and fling down bags of homework. A metal gong sounds, and the children gather up balls, skipping ropes and books and troop through a doorway into a large, roofless classroom. The Zhongdong Cave community has its own school, with 6 classes totalling nearly 200 children. Liu and eight others are weekly boarders, their families too far away for them to commute. So why live in a cave? The local government has repeatedly tried to persuade these latter-day troglodytes to resettle in purpose-built houses closer to civilization. But so

far, the community has resisted. They like their cave – it is warm in winter and cool in summer, is well ventilated and has a reliable supply of clean drinking water, good sanitation, mains electricity and satellite TV. It is home, and it works.

swifts and the high life

In Zhongdong Cave, fork-tailed swifts fly back and forth, swooping up into narrow crevices where their chicks perch in near-darkness on nests of fern-fronds held together with gummy saliva. The fork-tailed swift's relatives – the smaller cave swiftlets – are famous in Chinese cuisine as the source of that most bizarre of culinary delights, bird's-nest soup. The soup is a clear, glutinous concoction with little taste of its own – flavours are added – but that is not the point. It is the *texture* rather than the taste which is so admired by the Chinese gourmet – so much so, that the native species with the most prized nest, the Indochinese swiftlet, has been virtually exterminated in China by overcollecting of its nests.

Most of the trade is now outsourced to Malaysia, where cave swiftlets are exploited by criminal gangs, who stand to make a fortune from a natural commodity that trades for more than its weight in gold. Though fork-tailed swifts use little saliva in the construction of their nests, the price nests command offers enough incentive for a few individuals to risk their lives harvesting nests.

Over the hill from Zhongdong Cave is Getuhe Cave, owned by a local tourist consortium. Steps lead down from the riverside ticket office to the dock and a rank of narrow metal launches with cushioned seats and personal lifejackets. The boatman guns up the outboard engine, and you speed away downstream past feathery bamboo groves towards an awesome cliff. Up to the left is a huge tunnel that punches right through the limestone hill, admitting a dazzling shaft of early-morning sunlight. Silhouetted in the sun's rays are thousands of wheeling swifts.

As you approach the cliff base, a 60-metre-high (197-foot) cave-entrance comes into view, into which the river disappears. You can just about hear the roaring of distant rapids above the sound of the engine. There are even more swifts in this lower entrance – perhaps tens of thousands. As you enter the cave, the boatman cuts the engine to a gentle purr, and the echoing roar of the rapids fills the air. There are rocks ahead and a sandy beach, illuminated from a great skylight in the cave ceiling far above. You are assaulted by the pungent ammonia smell of guano – dark pellets of insect remains litter the sandbanks, and the air is filled with a drizzle of fresh droppings. You are glad you wore your Chinese plastic rain-cape after all.

TOP
Filming house swifts exiting from the entrance to Huishui Swift Cave, Guizhou.

ABOVE
A house swift resting on rocks by the entrance to the cave.

A voice hails you from above. The boatman returns the greeting. A man in cheap tennis shoes, jeans and a loud yellow T-shirt is nonchalantly clambering about on the greasy vertical wall 20 metres (65 feet) above your head. He has no safety equipment. The boatman makes the introductions: 'This is the spiderman. If you want, he can go right up to the roof.' It turns out spiderman once made a living from collecting birds' nests, but the practice is now outlawed under government conservation laws, and so he is using his skills and familiarity with the cave to earn tips from tourists.

It is July, and the newly fledged house swifts are trying out their wings, speeding around the cave entrance in a noisy gang. From on high they are watched by a pair of peregrine falcons perched on trees on the cliff-top. As the young swifts emerge into full sunlight, they are temporarily blinded. This is the danger zone. The male peregrine has taken off and is skimming above the cave entrance. He folds his wings and hurtles down, slicing through a group of swifts, pulling up sharply over the river and then powering back up to the cliff-top, his victim twitching as powerful talons squeeze the life out of it. The larger female on her perch stretches her wings, keening loudly as her mate flies by.

Half a day's drive to the east is another river cave that is home to an even bigger colony of house swifts – an estimated quarter of a million birds. Local villagers visit regularly to sweep up sack-loads of guano to use as fertilizer on their vegetables. They hope one day to develop their cave as a tourist attraction and are keen to guide visitors into its dark recesses. One draughty passage continues high above the river to a series of crystal dams that slope up into a large, dry, domed chamber inhabited by bats. The air is surprisingly cold until about halfway up the chamber, when it suddenly gets warmer. The boundary between the cold air below and the warm air above is marked by a layer of cloud – it's a cave with its own internal weather system.

With Chinese cave exploration still in its infancy, who knows what marvels may yet be discovered beneath the vast karst landscapes of southern China. Yet the subterranean world is only a small part of this amazing landscape.

the monkey show

The ruggedness of the karst has in some cases protected its original forest cover, despite a massive acceleration in the felling of forests from the mid-1950s to the mid-1970s to fuel China's rapid industrialization. As one of China's poorest and least developed provinces, Guizhou was slower to respond and has therefore retained rather more of its native vegetation and wildlife.

The remote Mayanghe area in northwest Guizhou, close to the Hunan border, is particularly rugged, with great karst gorges and impossibly steep, forested hills. This is home to a population of one of China's most engaging primates – the long-limbed Francois' langur, also known as the black leaf monkey. Adults are jet black with pointed, conical head tufts and white handlebar moustaches; infants up to the age of

about three months are bright orange. Until a few decades ago, the monkey was widespread throughout the karst of southern China. But having been hunted intensively for meat and for the supposed medicinal properties of its bones, which were until recently made into a tonic known as monkey wine, only tiny scattered populations remain. Selling monkey wine is now illegal, and the monkeys are protected by law.

The best time to see the langurs is in the late evening, when a troop comes together in preparation for bedtime. You hear them before you see them – rustling branches giving away the location where half a dozen monkeys are feeding among the treetops. They move quickly, picking off tasty buds and shoots from the extremities of slender branches, using their long tails for balance – the youngsters scrapping and tumbling yet somehow always managing to stay one hand-grab away from disaster. As the light fades, the monkeys transfer one by one to the rock face of a cliff. You cannot imagine what holds they can be finding as they move quickly and confidently up the near-vertical face – or why they should be risking such danger.

This section of the cliff is decorated with vertical brown stains – marking well-used routes. The infants whose rock-climbing skills are not yet up to scratch cling to their mums' fur as the troop moves higher still, towards a shallow cave. The light is almost gone as the last monkey disappears from view. Each troop of leaf monkeys

has two or three favourite caves – always on steep cliffs – which they use as night shelters. It seems to be behaviour deeply rooted in the monkeys' culture and may hark back to a time when nocturnal predators such as clouded leopards were common in the forests of southern China.

For anyone used to humid subtropical woodlands in other parts of the world, it is not just the absence of predators or larger mammals that marks China's forests as different but also the general scarcity of any animals – even birds and insects. The principal reason is the presence everywhere of human hunters. But hunting is just a sideline for the local communities. What sustains people – in astonishing numbers, considering the ruggedness of the landscape – is agriculture.

rice for the people

Virtually every hectare of cultivatable land in southern China is given over to food production. The slopes of dry, rocky hills may have single maize plants growing in pockets of soil, clumps of chilli bushes and carefully contrived earth terraces growing

THE CARP PONDS

Many paddy fields yield a second, equally important harvest: carp. The connection between carp and rice may go back to the earliest days of rice cultivation, as several species of wild carp are naturally found in shallow ponds in southern China. Theirs is a mutual association which benefits both parties: the herbivorous carp graze water plants, so helping to keep the paddies weed-free, while the rice plants shelter the growing fry from predators. Four species of carp predominate – bighead, silver, common and grass – providing both sustenance and cash.

the connection between carp and rice may go back to the earliest days of rice cultivation, as several species of wild carp are naturally found in shallow ponds

sweet potatoes and squashes. Houses, too, are often built on slopes, partly because the rock provides sound foundations and good drainage but more significantly because the flat, fertile land in the valley bottoms is too precious to cover in buildings. Such land is largely given over to the growing of just one crop – rice.

Historic records of rice cultivation date back 4000 years, and in classical Chinese, the words for agriculture and for rice culture are the same, indicating that rice was already the staple crop at the time the language was taking form. Rice is a member of the grass family, with wild varieties still found in the hills of Yunnan and elsewhere in southern China. The main species of long-grained rice cultivated throughout Asia today – *Oryza sativa* – traces its ancestry to the Yunnan species *O. rufipogon*, a wild perennial, which gave rise to a wild annual, ending up as *O. sativa* – a cultivated annual entirely dependent on human propagation. Archaeologists have found preserved grains of both wild and domesticated rice among pottery and other human artefacts in cave sites in Jiangxi and Hunan provinces that are 10,000 years old, suggesting that rice has to have been domesticated in Yunnan even earlier.

Piecing together the origins of such an ancient rice-growing culture can never be an exact science, but most scholars agree that cultivation of wild prototypes preceded domestication. It seems likely that rice grains were gathered by prehistoric people living in areas of high rainfall, where the wild perennial plants grew on poorly drained sites. These people also hunted, fished and gathered other edible plant parts. Eventually, however, their liking for the easily cooked and tasty rice grains led them to search out varieties that bore larger panicles and heavier grains and to transplant

BELOW
Bai women doing the back-breaking work of
planting rice in the paddies. Flooding the fields
helps the rice grow fast and keeps down weeds.

in classical Mandarin, the words for agriculture and for rice culture
are the same, indicating that rice was already the staple crop at the time
the language was taking form

them in boggy places closer to their dwellings. They would also have selected earlier-maturing, short-grained varieties that would grow on dry hillsides and hilltops. Over a long period, selective local cultivation resulted in a diversity of regional strains of rice suited to specific conditions. Even today, there are reckoned to be in the order of 140,000 distinguishable varieties.

As rice culture spread further north and east into areas of China where winter temperatures fell below freezing, the cultivated forms became true domesticates, depending on human care for their perpetuation. In parallel, the water buffalo – whose wild relatives still live in the swamps of India and Burma – was brought from the swamps of the south into the north and evolved as another domesticate.

China today is the world's largest producer of rice, at about 180 million tons per year, almost all of which is grown in the south. Nowhere better shows the impact of rice-growing on the landscape than the Yuanyang area of southern Yunnan, where many hundreds of terraces cover a series of hillsides 2000 metres (6560 feet) high. When you consider that each field has been carved from the hill slope and levelled using basic hand tools, the sheer scale of the effort involved cannot fail to impress – matching the Great Wall both in its ambition and in its execution.

A typical southern Chinese greeting goes: 'Have you had your rice today?', to which one is expected to reply enthusiastically: 'Yes!' In Yunnan, the saying is: 'Between the heaven and the earth, rice possesses the supreme position.' Young girls

ABOVE
Traditional Miao houses near Leishan, Guizhou, stacked up the hillside, with vegetable gardens and rice paddies packed in between. Animals are kept on the ground floor and rice is stored in the roof area.

with finicky appetites are warned that every grain of rice they leave in their rice bowl will be a pock-mark on the face of their future husband.

Rice cultivation is very labour-intensive and requires plenty of water for irrigation – on average, it takes 2000 litres (528 gallons) to produce a kilogram of rice. The irrigation systems required to manage such quantities of water are sophisticated, requiring the cooperation of hundreds of paddy owners. This has moulded the culture of Chinese rural communities. As a visitor, you can't help but be charmed by the spirit of cooperation that pervades such communities – and nowhere more so than among the Miao people of Leishan county in Guizhou.

the swallow forecast

The Miao traditionally live in villages of jam-packed wooden houses with heavily tiled roofs, which are stacked up the steepest and least productive hillsides, leaving every inch of cultivatable valley land free for growing rice and vegetables. As with most rice-growing cultures, the busiest times are the planting and harvesting seasons, when everyone has to work together to refurbish and prepare the paddies, plant the seedlings and then bring in the harvest. Leishan is a hilly area typical of Guizhou, scornfully dismissed in the popular epithet: 'endless hills, sunless skies and penniless people'. The

BELOW
A red-rumped swallow – harbinger of spring – resting on a man-made perch in the window of a traditional Miao house.

winters are cold, and the timing of spring is unpredictable. But the Miao have hit on a clever way of forecasting what the coming season has in store. It's all to do with swallows.

Each home follows a similar layout, with a large front room facing the rice paddies. Come spring, at least one window is left open, and attached to the ceiling beams of the front room are wooden brackets – bases for the swallows to nest on. By mid-April, most homes are occupied by one or more pairs of red-rumped swallows, coming and going with beakfuls of mud to refurbish last year's nest.

The Miao regard the swallows as a sign of good fortune and of marital bliss – they believe the birds pair for life and that the same pair returns each year to its home. They

also believe the swallows can forecast the seasons, and use the date of the swallows' arrival to predict the optimum time for planting rice. Once the wise man has determined when this should be, a cooperative planting plan is prepared. The resulting transformation is a magical sight. A gang of villagers can fill a half-acre paddy with evenly spaced rice seedlings in a couple of hours. Within three or four days, the entire patchwork landscape metamorphoses from muddy brown to emerald green. Then the celebrations begin.

vanishing beasts

Fish-eaters such as otters, ospreys and fish eagles, as well as rice-eating birds such as sparrows, buntings, munias and avadavats have been persecuted for centuries. Add in forest clearance – for firewood and construction – and the culture of consuming wild animals as food or medicine, and the scarcity of any wildlife other than species regarded as friendly or of cultural significance is not so surprising.

Some southern Chinese animals such as the great one-horned rhinoceros and the giant panda vanished long ago. Others such as the south China tiger and black bear have disappeared recently. Many others continue to survive as extreme rarities, including predators such as the leopard, the larger reptiles, most of China's freshwater turtles and a number of amphibians, including the Chinese giant salamander – the world's largest amphibian.

The Chinese giant salamander resembles a huge, blotchy, grey cucumber fringed with flaps of skin and covered in warty bumps and slime, with stubby legs growing out of its sides. It has a flattened head, a wide trapdoor mouth and tiny piggy eyes and may grow to more than 1.5 metres (5 feet) long. An ambush predator, it relies on near-perfect camouflage to hide on the floor of shallow streams, where it gulps down any passing fish or other mouthful. The Chinese call it the 'baby fish', because apparently it can utter a call that sounds like a crying infant. This monstrous creature was once found throughout southern and central China – particularly in the karst – but is now rare. The main reason is that it apparently tastes delicious and was highly prized as a centerpiece at important banquets.

canyons and cascades

One area where the giant salamander allegedly still exists is in the streams of remote canyons in the Zhangjiajie area of Hunan province. The fabulous rocky landscape of the Wulingyuan Scenic Area north of Zhangjiajie is perhaps the most dramatic in all China, with slender stone pinnacles soaring up to 300 metres (985 feet) from narrow wooded canyons filled with tumbling cascades.

These unique geological formations look like an extreme form of karst but are formed of hard quartzitic sandstone. No one has yet come up with a convincing explanation for why the Wulingyuan pinnacles have such extreme shapes, but they are now one of China's most popular scenic attractions, drawing more than 2 million tourists a year.

South of Zhangjiajie, a number of rivers flowing eastwards out of the karst country unite to form the Yuan Jiang, which flows eastwards into the great Dongting Lake – part of a vast complex of abandoned oxbows and tributaries of the Chang Jiang, or Yangtze, River. This marks the eastern end of a vast stretch of lowlands known as the Land of Fish and Rice.

the great lakes

The Land of Fish and Rice presents a completely different face of rural China from that of the rocky hill country to the west. This is a more affluent landscape, with a far denser human population. There are roads and power lines everywhere, and agricultural production is far more mechanized. Given the scarcity of wild animals in the less-populated karst landscapes, what hope could there be for *any* to survive in such a crowded place as this? Yet this is an area of crucial importance to wildlife.

Dongting Lake once covered a vast area of about 4300 square kilometres (1660 square miles), but increased demand for rice starting in the Liu Song dynasty (the mid-fifth century AD) resulted in a steady reduction in its size, as land was reclaimed for agriculture by building dykes. Land drainage along the middle and lower reaches of the Yangtze accelerated between 1950 and 1980, when an estimated 12,000 square kilometres (4630 square miles) of lakes and tidelands were claimed for agriculture, causing the lake to shrink and greatly reducing its population of fish and birds.

Then in 1990, the government changed its policy, returning a large area of farmland to lake, relocating thousands of people and compensating them for the loss of their paddy fields. At the same time, many bird species were afforded strict government protection. Fish stocks increased, the birds returned, and the lake now provides a crucial haven for wintering flocks of rare species such as swan goose, Bewick's swan, oriental stork, and Siberian, hooded and white-naped cranes.

Further east, the even larger Poyang Lake is another crucial winter refuge – with

LEFT
The rarest crane in the world – the Siberian crane. The majority of the surviving birds winter in the vitally important wetlands of Poyang and Dongting, and then migrate back north to breed in Siberia.

an estimated 40,000–50,000 swan geese, 60,000–70,000 Bewick's swans and about 3000 each of the endangered oriental stork and critically endangered Siberian crane. The latter species is known in China as the white crane, and Poyang and Dongting lakes provide two of its three winter refuges (the third being in India), which between them shelter more than 95 per cent of the world population.

Standing 1.4 metres (about 4.6 feet) high, with pure white plumage, black bill, yellow eyes and vivid red face and legs, the Siberian crane is strikingly beautiful, with a haunting, flute-like call. Though strictly protected in both their summer breeding grounds in northern Russia and their Chinese wintering grounds, the migrating cranes run a gauntlet between the two, and so are wary of humans. At Poyang Lake, it is almost impossible to approach closer than about 200 metres (655 feet) before they fly off. Other birds at Poyang are equally shy, even the large flocks of lesser white-fronted and swan geese. Given that the reserve area used by the birds is more than 220 square kilometres (85 square miles) – the entire lake covers 5000 square kilometres (1930 square miles) – the key to successful birding is to allow plenty of time and come armed with a powerful telescope.

Though conservation areas are now well established in several wetland areas of the middle and lower Yangtze River basin, there is still a severe problem of disturbance by local fishermen plying the lakes in noisy motor boats, making it hard for the birds to feed undisturbed and to build up their reserves during the critical preparation for their spring migration and the coming breeding season.

muddy dragons

East of Poyang Lake, the Yangtze River leaves Hubei province, entering Anhui province at its southwestern tip. From here it flows diagonally northeast towards Jiangsu province, before swinging back southeastwards past Shanghai, to empty into the East China Sea. The countryside of southern Anhui is a maze of small rivers, shed by the mountains to the south. Among the flooded rice-paddies, Chinese pond herons patrol in search of frogs, while pied kingfishers hover like hummingbirds over the limpid rivers.

It is early summer in the Land of Fish and Rice, and something is making the ducks nervous. A female mallard rounds up her chicks. A coot alarms noisily in the reeds. Close to the bank, a dark head pops up. Reptilian eyes survey the waterside vegetation, and the floating head turns to reveal a wide snout edged with downward-pointing conical teeth.

The Chinese alligator – known in China as tu long, or muddy dragon – is the only alligator found in the Old World. How it came to be here is a historical and geographical puzzle. Alligators belong to a very ancient lineage, the archosaurs, which pre-date the dinosaurs. By 140 million years ago, alligators in the New World

BELOW
China's alligator, the tu long, now very rare in the wild, though thriving in captivity.

BELOW
A large flock of pied avocets on Poyang Lake, the largest lake in China in the wet season. Siberian cranes are the main attraction, but other stars include oriental storks, swan geese, and white-naped and hooded cranes.

there is still a severe problem of disturbance by local fishermen plying the lakes in noisy motor boats, making it hard for the birds to feed undisturbed and to build up their reserves

~

THE DRAGON KEEPER

'I call them all Dragon Chang, because to me they are like my own sons and daughters. Everyone around here knows me, and they know not to harm my children'

'You are going to meet Dragon Chang,' says Chang Jinrong, holding a plucked chicken in tongs. A pair of broad, toothy jaws emerges from the pond, accepts the fowl with surprising delicacy and gulps it down. 'There are nine alligators here,' says Chang, who is in his late seventies. 'I call them all Dragon Chang, because to me they are like my own sons and daughters. Everyone around here knows me, and they know not to harm my children.'

Chang loves the notoriety that comes from his association with the alligators in the ponds around his home. The walls of the little hut-cum-guard-post where he hangs out most days are decorated with painted slogans – Protect the Chinese Alligators – and with press cuttings showing him feeding his charges. 'I am famous in all this region,' he announces proudly.

Chang's dragons are completely wild – they just happen to live in a heavily populated agricultural part of southern Anhui province, surrounded by villages, noisy roads and a jungle of electrical pylons. Unfortunately, the reptiles have the habit of digging deep holes around the edges of the water bodies where they live, sometimes undermining rice-paddy earthen dams, which are critically important in controlling water levels during the growing season. Faced with general hostility throughout their range, it is only people such as Chang who stand between the muddy dragons and extinction.

were developing their trademark broad skull, with the fourth upper tooth fitting into a socket in the lower jaw, concealing it when the jaw is shut. A surprising number of species of alligators have been found in fossilized rocks in various parts of North America dating from the Cretaceous period, but not in China. Scientists speculate that, during the late Cretaceous, distant ancestors of the Chinese alligator spread into eastern Russia across the land bridge that joined America and Asia where the Bering Strait now is. Having reached Asia, they eventually spread south into China, where they evolved into the tough little survivor we know today.

Compared to their American cousins, Chinese alligators are surprisingly small – seldom exceeding 2 metres (6.5 feet) long. They feed on birds, fish, frogs and invertebrates, which are plentiful throughout the Yangtze River floodplain and its countless lakes, swamps and tributaries. The climate here is relatively mild, with summer highs about 35°C (95°F) and winter lows seldom much below freezing. Like most cold-blooded reptiles, alligators avoid temperatures below about 10°C (50°F) by hiding in underground burrows. Hibernation begins about the end of October, and the adults emerge again in late March. By mid-May, they are ready to begin their courtship – the males bellowing loudly to attract mates.

These days, there are perhaps only 150 Chinese alligators left in the wild, making this one of the world's most critically endangered crocodilians. The causes are all too familiar: persecution (because their digging activities around paddy fields and their appetite for ducks and fish brand them as pests) and hunting for the pot and for their supposed medicinal properties, including the ability to prolong life. Only the strictest government protection ensures their continued survival.

The miracle is that any remained at all by the time the Chinese government began to take an interest in their welfare in the early 1980s. The previous decades had been a period when China's human population had more than trebled, resulting in much land previously considered not worth farming being pressed into

BELOW
A bridge linking two huge granite pinnacles of Huang Shan – the Yellow Mountains – a World Heritage Site famed for its scenic beauty and its ancient Huangshan pines. Many of the plant species growing on the mountains are found nowhere else.

cultivation. Wildlife surveys conducted in the late 1970s alerted the authorities to the crisis facing their unique 'muddy dragon', and in 1982, the government's forestry department began a project to establish reserves to protect the few remaining tiny populations. At the last count, alligators are spread between 13 ponds in an area the size of a typical English county.

On the outskirts of Xuancheng in southern Anhui province, a captive-breeding centre has also been set up, which houses about 500 adult alligators – the oldest more than 40 years old. Each year, about 1600 eggs are collected from nests constructed by the breeding females on islands and are then incubated. No young alligators are being released from the centre, supposedly to avoid disturbing the genetic make-up of local wild populations. Instead, surplus alligators are sold by the centimetre and feature on the menus of local restaurants.

the riches of caohai

Alligators aren't the only dragons in the Land of Rice and Fish. During the summer months, the waterside vegetation of every southern Chinese stream, pond and marsh is alive with spectacular dragonflies and demoiselles. There are huge golden-banded hawkers and orange and sky-blue-bodied chasers, plus a host of smaller, jewel-like damselflies, some species forming huge swarms.

In Guizhou province, one little-known wetland perching at 2200 metres (7200 feet) among spectacular mountain scenery, close to the border with northern Yunnan, is so rich in dragonflies that they form the basis of a local fishery. In summer, the villagers of Caohai harvest the plump dragonfly nymphs of *Anax partheope* using fyke nets and spread them out to dry on the flat roofs of their homes. Sacks of the dried insects are dispatched by local traders to restaurants across southern China, where they are served up as a culinary delicacy.

There are other mini-monsters at Caohai. The rare Guizhou crocodile salamander, also known as the red-tailed knobby newt – a pugnacious, stubby-legged, warty-skinned predator with an outsize mouth filled with tiny, needle-sharp teeth – is found only here. Little is known of its behaviour in the wild, but it is thought to feed mainly on worms, insects and other land invertebrates. When

threatened, it produces a poisonous secretion from its warty skin as a defence, and its vivid scarlet-and-black colour – designed as a warning to predators – also qualifies it as one of the most strikingly beautiful amphibians in China, making it highly sought after by collectors.

Caohai – literally 'sea of grass' in Chinese – is rated by WWF as one of the top-ten birdwatching destinations in the world and includes a forest nature reserve designated specifically for the protection of the rare endemic Reeves' pheasant. But it is the marshy wetland which is the principal attraction for birdwatchers. In the winter, the area hosts a spectacular influx of up to 100,000 birds from the Tibetan plateau and beyond, with 179 species recorded, including Eurasian and hooded cranes, oriental stork, black stork, bar-headed goose and ruddy shelduck. But the undoubted star of the show is a population of close to a thousand black-necked cranes – the world's last remaining high-plateau crane.

monkey valley

The streams which sustain the rice-lands and alligator-ponds of Anhui province originate from the great granite massif of Huang Shan to the south. Huang Shan (Chinese for yellow mountain) is famous throughout China for two things: its fabulous granite pinnacles, which seem to float above a sea of clouds, and its contorted ancient pine trees, immortalized in endless classical Chinese paintings. Like Guilin and Zhangjiajie, Huang Shan is a magnet for Chinese tourists and photographers, who are prepared to queue for hours at one of three cable-car stations for a chance to shiver in the chilly mountain mist. Surprisingly few bother to make a small detour into one of the valleys at the base of the mountain, where a far more rewarding experience awaits – at least for those interested in wildlife.

Monkey Valley is home to a subspecies of the roughest, toughest, meanest primate in China – the Tibetan macaque. Though relatively short and squat, male Huang Shan macaques may weigh up to 12 kilos (26 pounds) and are immensely strong and frequently aggressive. Any intrusion into their territory requires some caution and the services of a knowledgeable guide familiar with the rules of monkey etiquette. The first rule is to show respect at all times by avoiding rude stares or direct eye contact. To flout this rule is to invite trouble. Any monkey that catches your eye will almost certainly launch a full-on threat display, involving a furious and terrifying charge. Most local guides carry a large stick or a handful of rocks to dissuade such overt aggression. But if proper care is taken, an hour or two spent with the macaques can be a fascinating experience.

Both sexes are covered in luxuriant brindled grey-brown fur that looks fabulously warm – perfectly suited to the mountain weather. The males are the more powerfully built sex, with a face reminiscent of an adult male orang-utan's – small, close-spaced eyes set in a wide, disc-shaped face. The less bulky females are distinguishable from a distance by their purple-pink eyelids, which they use to signal submission or displeasure or simply to flirt. When not feeding or travelling, the youngsters seem to spend every available minute fighting, biting and pulling each other's tails in great tumbling heaps, providing endless entertainment.

the sucker frog and the hundred-pace viper

The streamside where the monkeys gather several times each day to be fed with handouts of corn by reserve personnel is home to another very special creature. The concave-eared torrent frog, also known as the Anhui sucker frog, has two unique attributes. It is the only frog with an ear canal and the first non-mammalian vertebrate known to produce an ultrasonic call – of a higher pitch to that used by most bats. Scientists believe that the two phenomena are linked – the frogs' thin, recessed eardrums are more sensitive to high frequencies, allowing them to hear each other's ultrasonic calls over the lower-frequency white noise of the rushing water torrents where they live.

Until recent times torrent frogs may have fallen prey to a voracious predator of the local streams – the strikingly coloured golden-headed box turtle – one of China's 25 species of native freshwater turtles. Turtles in China are hugely sought-after for food and medicine, and most are endangered or ecologically extinct – including all eight box turtle species, individuals of which may trade for up to $200,000 each.

But one local reptile seems to be holding its own in the area. The hundred-pace viper – also known as the five-pace viper, or Chinese moccasin – is considered the most dangerous snake in China. It is a spectacular reptile, growing up to 1.5 metres (5 feet) long, with a thick, muscular body, patterned a little like a diamondback rattlesnake, and a wide, triangular head with an upturned, pointed scale at the tip of its nose. The local macaques are terrified of it and have a distinctive alarm call to warn each other the moment a snake is spotted so they can make a quick exit into the nearest tree. Fortunately, the viper is a ground predator, specializing in ambushing rodents and birds – mainly at night and often close to water. For now, its cryptic colour and mountain-forest habitat have safeguarded it from overexploitation, though it remains popular as an ingredient in medicinal 'snake wine'.

ABOVE
A hundred-pace viper, feared by monkeys and prized by humans as an ingredient in medicinal wine.

a future for the wild?

The outlook for the wildlife of south China would seem to be mixed. There are encouraging signs that the Chinese government is taking steps to create an ever-increasing number of wildlife reserves and has outlawed trade in many wild species – though enforcement is extremely difficult. The real cause for concern is the persistence of the widespread culturally ingrained belief among some people in southern China that nature exists primarily to be used by people, as food and medicine. As the rapid increase in disposable income among city-dwellers fuels an increasing demand for wild food – considered to be particularly good for health – the future of southern China's wildlife has to remain a cause for concern.

6

crowded
shores

MUCH OF CHINA'S MOST SOUGHT-AFTER LAND lies along the 14,500km-long (9010-mile) coast and its islands. So the wildlife that survives here has to jostle for space with people, more so than in any other part of China. Intriguing relationships have evolved from this close association, but to understand China's coastal life, you first need to look at the past.

The outline of China today owes its origins to the rising sea levels that started around 12,000 years ago, caused by a warm period towards the end of the last ice age. A survivor from those dramatic times can be found in China's northernmost Bo Hai Gulf, where sea ice still forms in winter. The spotted seal is the only seal to breed in Chinese waters, hauling out on the ice in spring to give birth. Once their pups are able to catch fish for themselves, the seals gather to the north of Bo Hai and then head south to the Shandong peninsula, where they moult, spending the summer off the coast of South Korea and North Korea.

Hunted for their coats, flesh and penises, the seals are nervous, and it is hard to get near them by boat. Though they are protected in Chinese waters, spotted seals still face threats of poaching and pollution.

ABOVE
Spotted seals off the coast of northern China. They are the only seal to breed in Chinese waters, but they are still sometimes hunted, illegally, and so are very wary of people.

PREVIOUS PAGES
An undeveloped beach in Fujian province. Flat coastal land is increasingly sought after for tourist, agricultural or industrial development.

Another animal affected by the change in sea level at the end of the past ice age is the Shedao pit viper. Like many animals that didn't run, swim or fly away when the sea level rose and hills became islands, they were stranded. On Shedao Island they were forced to change from a diet of rodents to one of small migratory birds and the odd insect or other invertebrate, resulting in a feast-and-famine regime. For ten months, the snakes have little or no food and lie dormant in their burrows, but in autumn and spring, Shedao's freshwater pools attract an influx of tired and thirsty songbirds travelling south to Australasia to avoid the worst of the northern winters and then back north to Siberia in spring to breed. The commotion of their arrival, combined with increasing temperatures, triggers the snakes into action, and they climb up the shrubs and lie in wait, often returning to favoured ambush sites.

Incredible timing and accuracy are needed for a successful catch, as the snakes only have a short strike-range. Even when a bird is caught, its last struggle may cause the snake to lose its grip and its meal to a viper below. Today, there are so many vipers on the small island that they form one of the densest populations of snakes in the world.

BELOW
A Shedao pit viper warming up in the sun before climbing a branch to lie in wait for passing birds.

the birth of civilization

The great rivers of China flow from an icy hub on the Tibetan Plateau. When they reach the sea, the Yellow River (Huang He), Yangtze River (Chang Jiang – historically, Da Jiang) and Pearl River (Xi Jiang) form estuarine wetlands that are crucial for much of China's coastal wildlife.

The mother river – the Yellow River – crosses nine provinces and travels about 5500km (3420 miles) before emptying into the Bo Hai Gulf. But this hasn't always been the case. The river's heavy load of sediment has resulted in frequent changes of course, and at times it has emerged to the south of the Shandong peninsula, making this delta one of the fastest changing areas of coast on the planet.

This dynamic area became one of China's earliest sites of civilization. The rich sediments of the delta and the associated abundant marine life attracted the Shao Hao tribe – one of the first to change from a life of hunting and gathering to one of agriculture. Though these people lived on the delta, they were drawn inland to Mount Jinping on the fertile plateau. Here, 7000 years ago, on the rounded granite surface of the Jiangjun cliff, they recorded their lives. Their carvings include human-like figures linked by vertical lines to sheaves of wheat – the first illustrations of agriculture in China. But perhaps the most stunning petroglyph is that of the Milky Way, lending magic to this ancient site. One can only marvel at the time and patience needed to create China's first star atlas. It is thought that this rock was used as both an astronomical observation point and an altar for phallism and ancestor worship.

The Yellow River has been engineered by humans for millennia, but with 120 million people now directly dependent on its water, extraction and diversion have intensified to such an extent that, today, little water reaches the coast. In 1997, the delta was dry for 226 days. Records show that the river has created 2708 square kilometres (1045 square miles) of wetland along the coast since 1855, but this has now started to reduce in size and is shrinking southwards.

THE RED-CROWNED CRANES

Part of the vast freshwater wetlands of Heilongjiang province are protected as the Zhalong Nature Reserve. Here red-crowned cranes arrive in spring to breed. To witness a dawn duet, when a pair reaffirms its long-term partnership and stakes claim to its nest site, requires getting up before dawn and wading through ice-covered water. But at the reserve's captive-breeding centre, it is possible to get a closer view. Here, eggs are taken from nesting cranes early in the season, to allow time for them to lay a second batch, and incubated by a husband-and-wife team. The couple take shifts through the night to turn the eggs and assist at the birth of the long-legged chicks, which are so large that it's hard to believe they ever fitted into their shells. The chicks are hand-reared, and tourists gather daily to watch the adolescents exercising their wings.

Some of the released birds have raised their own chicks in the reserve. But the real challenge for the wild birds is to survive the migration south, fly-hopping down the coast between salt marshes and deltas. Their first stop-off is the Shuangtai estuary of the Liao River at Liaodong Bay to the northeast of the Bo Hai Gulf – one of the largest continuous reedbeds in the world. They join another 106 species of birds refuelling here before flying south once more to the Yellow River estuary and then down to Yancheng Biosphere Reserve north of Shanghai.

eggs are taken from nesting cranes early in the season, allowing time for them to lay a second batch, and incubated by a dedicated husband-and-wife team

delta wetlands

China's 2 million hectares (5 million acres) of tidal flats have been used for reed cultivation and aquaculture – prawns, clams, mussels and oysters – and for growing rice and wheat for many thousands of years. Sea-salt production has also been an industry along the coast since 119 BC, and today the tidal flats produce more than 20 million tons of salt a year. As a result, little coastal land is left for nature.

One site that is protected is the Yancheng ('salt city') Nature Reserve on the coast of the Yellow Sea, where each autumn, waves of geese, ducks and cranes, including more than a third of the world's population of red-crowned cranes, fly in for the winter to feed on the saline meadows and mudflats. One of the most awesome spectacles is that of thousands of spot-billed ducks rising from the water with a roar of wings in an effort to evade swooping eastern marsh harriers and creating spectacular patterns in the sky as they wheel and whirl like flocking starlings. Some mass flights, though, are unfortunately caused by illegal gunfire.

Yancheng and nearby Dafeng Nature Reserve are home to a rare and skittish deer. Extinct in the wild in China since the last one was killed in the early 1900s, the

BELOW
Père David's deer at the coastal wetland of Dafeng Nature Reserve. The deer's feet are splayed to help support their weight on boggy ground.

sze-pu-shiang, or Père David's deer, is now thriving on the protected coastal grassland. Often referred to as the milu (though milu means sika deer), its fate has been linked with humans since prehistoric times, when it was hunted in the coastal wetlands.

The first foreigner to look over the wall and into the emperor's imperial hunting park of Nan-Hai-Tze, just south of Beijing, was the French missionary and explorer Father (Père) Armand David in 1865. Seeing a herd of milu, he realized that the species was unknown to western science, and he later managed to smuggle out two skins and antlers. After several attempts at exporting the deer, the first pair of live ones were sent to Europe in 1869, and in 1898, the 11th Duke of Bedford began to breed them in his grounds at Woburn Abbey, just before the last of the wild deer were shot. The Woburn herd prospered, and in the 1980s, 40 were returned to China. These bred well, and a recent count put the number of deer living in three Chinese national nature reserves at 2500.

Sze-pu-shiang means 'none of four'– a Père David's deer is said to have the feet of an ox, the neck of a camel, the antlers of a deer and the tail of a donkey. Such attributes have a purpose. Splayed feet support the deer's weight in boggy conditions, a long neck enables it to feed on deep-growing water weeds and the backward-pointing tines of its antlers keep them from getting caught in the dense reeds. The deer often swim across large stretches of water and have even been seen dozing in water on a hot day with their noses under water.

ABOVE
Red-crowned cranes calling in duet to proclaim ownership of their nest site at Zhalong Nature Reserve.

The males start rutting in the first two weeks of June, adorning their horns in grassy finery. The younger males play-fight, but serious fights involve a turn of speed and clash of antlers. The winning males inherit the females and their fawns, which gather in crèches that keep them safe from the aggressive males hounding their mothers.

rocky shores and sheltered bays

The majority of China's coastline is muddy, but a granite promontory, the Shandong peninsula, partly encloses the Bo Hai Gulf, its rocky shores pummelled by sediment-heavy waves. Naturally occurring bays provide havens for both wildlife and people and include China's oldest naval harbour, Penglai Water City. Close by, perched on Danya mountain, is Penglai Pavilion, a Song dynasty temple (960-1127). Penglai refers to a supernatural mountain supposedly on an island in the Bo Hai, inhabited by eight legendary Daoist immortals. It is said that the immortals got drunk at Penglai and crossed to the island without the use of boats, some on the backs of cranes. The mortals who built the naval port here were influenced by geography rather than mythology, as this is the point where the Bo Hai Gulf meets the Yellow Sea, making it important for both trade and protection.

Close to Penglai is the village of Chuwang, where a fishing festival takes place in the first month of the lunar calendar. Traditional rituals are performed to appease the goddess of the sea, so she will protect their ships and provide bountiful catches. As a boat owner, it is up to Mr Zhao to entertain and feed his crew and their families in a set of rituals and feasting. Following a huge meal, an enormous fish and a pig head are dressed in red ribbons and taken in a noisy procession down to the port. The route is strewn with firecrackers to such an extent that the air turns white and the sea red with firecracker paper. The offerings are then taken to the bow of the boat and prayers said, at which point, fireworks, dragon dancing and music breaks out in the harbour. As the quiet twilight hours arrive, a few people keep with tradition and send small handmade boats out to sea, each lit with a single red candle.

To provide marine food for an increasingly demanding population involves aquaculture on a huge scale. Molluscs such as oysters, mussels, abalone and scallops, together with seaweed, have always been harvested from rocky shores. In the 1950s, the Chinese started to use floating rafts in the calmer waters to grow molluscs and algae on ropes. Haidai, a species of kelp, is grown in vast amounts. It was originally brought over from Korea and Japan and has been cultivated on ropes in China since 1956. Kelp aquaculture has also inadvertently become a life-saver for the winter angels, better known as whooper swans, which for thousands of years have been migrating south to overwinter in four bays in the Yangtai area. The swans have lived here for so long that the daily routines of the people and birds are now interlinked.

Mr Qu, aged 86, and his daughter and son-in-law have lived here for most of their lives. In the winter, his daughter rises early and gets the fire going to warm the house before donning a colourful headscarf and setting out to the shore with a rake and bucket. She and her friends gather for a chat before getting down to the business of collecting cockles and razorshells for the day. In the distance the whooper swans doze – familiar with their movements, bright headscarves and chatter. As the women start to make their way home for breakfast, the swans stretch their wings and fly out to sea for their own repast. They fly over Mr Qu as he chugs out to the thousands of ropes hung from the buoys at the kelp farm. While he tends

BELOW
Mythical figurines on a roof at Penglai, the number denoting the importance of the site. The emperor placed 11 such figurines on roofs in the Forbidden City.

following a huge meal, an enormous fish and a pig head
are dressed in red ribbons and taken in a noisy
procession down to the port

~

the young kelp blades and checks the buoys and lines, the swans arrive to feed on seaweed known locally as sea grass, which grows naturally between the buoys.

This seaweed once grew on the rocky shoreline but no longer does so in large amounts, and the best place for the swans to find it is at the kelp farm. The villagers also miss the once-bountiful wild seaweed. They mostly live in traditional one-storey stone cottages that for centuries have been thatched with wild seaweed. A lack of this seaweed now means that thatching is a dying art.

When the wind picks up in the afternoon, workers and swans come back to shore, the swans to drink and bathe in fresh water, a rare commodity, and fights regularly break out for the best spots along the few streams. After a hard day's work in the sea wind, families gather round their central stove to steam bread and cook cockles, fish, rice and seaweed. The main heat from the stove is cleverly rerouted through pipes to warm their beds for the night. All that's left is for the younger villagers to take the spare bread down to the shore to feed the swans before turning in for the night.

The next rocky shores to the south do not appear until past Shanghai, from Hangzhou Bay to Guangdong and Guangxi. Fujian has the highest elevation of all the coastal provinces, with 90 per cent of its land made up of hills and mountains. Unlike rivers in the north, Fujian's rivers travel only a short distance from mountains to the sea. It is a hard land, regularly decimated by typhoons that storm into the coast, causing landslides and filling the rivers with raging water.

THE SILVER DRAGON

as one of the world's great wonders, it attracts thousands of tourists, who arrive on the eighteenth day of the eighth lunar month to see the spectacle

Qiantang bore, which occurs in Hangzhou Bay south of Shanghai, is a natural phenomenon triggered by the pull of the moon. When the tide comes in, a huge wave is produced that surges inland up the Qiantang River.

As one of the world's great wonders, it attracts thousands of tourists, who arrive on the eighteenth day of the eighth lunar month (September/October) to see the spectacle. The Chinese call it the Silver Dragon, and on a spring tide, the wave can be up to 9m (30ft) tall and reach speeds of around 27kmh (17 mph). In the twelfth and thirteenth centuries, suicidal surfers would ride the bore on small planks of wood, in an attempt to placate the dragon's wrath.

This is where the Hakka people live. Hakka is a Cantonese pronunciation of the Chinese words ko chia, meaning guest people. They are thought to have originated in northern China – Han court officials, forced to leave in several waves of migration, arriving in Fuijan some time between AD 317 and AD 907 . Their language, customs and traditions are among the oldest in China. Their folk songs are complex (though less prevalent since the Cultural Revolution) and include mountain songs – an impromptu form of sung conversations to describe their surroundings. During the tea harvest in April, all the women gather together to bring in the tea leaves. These are picked by hand or with shears on sunny days, when the flavour is better. Herds of goats neatly graze the weeds from among the tea bushes, leaving the bitter-tasting tea leaves alone but fertilizing them with their droppings.

There are three different types of Hakka homes. The phoenix houses were built in the style of the imperial court, meaning the families had the favour of the emperor, but as the Hakka fell out of favour and the locals began to attack them, the roundhouse became the design of preference, effectively forming fortified villages. When all became peaceful once more, flat-roofed houses were built.

The Hakka roundhouses were built on feng shui principles. Feng shui generally promotes harmony with the environment. The best orientation a building could

ABOVE
Buoys holding up ropes under which hang millions of farmed kelp fronds. The overwintering whooper swans feed on the wild sea grass seaweed that also grows on the ropes.

FUJIAN TEA

Tea has been drunk in China for 4000 years. The earliest written record of tea production in Fujian is on a stone tablet inscribed in AD 376. Tea still plays an important role in Chinese social life, it is offered to a guest entering a home as a way of showing respect, and serving a cup of tea is a symbol of togetherness.

The finest tea is usually grown at altitudes of 900–2100 metres (2955–6890 feet). The spring monsoon season (when the land warms up quicker than the sea, causing heat to rise and pull in moisture from the sea) provides the perfect damp conditions for tea. The most perfect cup of tea is made from the same rainwater that nourishes the plants or from spring water.

There are 336 varieties of tea in the province of Fujian, which makes it China's tea treasury. It produces all five types – black, green, white, oolong and scented – and all except green tea were pioneered by Fujian. Lapsang souchong (black) and white tea are Fujian specialities. Oolong tea, produced by the Hakka people among others, combines the zest and fragrance of both green and black tea. Oolong means black dragon, possibly because the leaves look like small black dragons when hot water is poured on them. After the tea is harvested, it goes through numerous drying, bruising, sorting and twisting processes before being packed ready for sale.

> the most perfect
> cup of tea is
> made from the
> same rainwater
> that nourishes
> the plants
> or from
> spring water

have is with its southern face to a river or a lake and its back to a hill in the north. Each roundhouse was built to outlast sieges and was fortified with thick mud-brick walls, with only a few high windows. It would usually shelter about 20 families – upwards of 100 people. Most roundhouses have 3–5 storeys and 30–60 rooms and are built around a circular courtyard – to let in light and provide space for a well and room for farm animals. The ground floor has kitchens and dining rooms, the first floor food storage, and the second floor bedrooms. The rooms on each level are the same size and open out onto a circular walkway. One can only marvel at the carpentry and engineering skills used to create high-rise buildings that have survived up to 800 years of earthquakes, sieges and typhoons.

tropical islands

The third type of shore habitat in China is made from living things – mangroves and corals. These occur south from Fujian, and on the islands of Hong Kong and Hainan, and islands in the South China Sea. Hainan province officially includes some 200 islands, including James Shoal in the Spratly Islands, which the People's Republic of China claims as its southernmost border at 40 degrees south, though ownership of many such islands is disputed with other countries.

Far off the coast in the south is the Kuroshiro warm-water current – the second largest in the world after the Gulf Stream – which flows northeast from the South China Sea towards Taiwan and Japan and allows coral to grow on the Japanese coast

ABOVE
Traditional tea-picking in a remote Hakka village in Fujian. Tea is mostly picked by the women and organized and processed by the men. Goats graze the weeds between the rows of tea bushes.

BELOW
Hakka roundhouses in Yongding county,
Fujian province. They are still lived in, though
mainly by the older generation, each family
with its own kitchen, storeroom and bedrooms.

most roundhouses have 3–5 storeys and 30–60 rooms and are built around a circular courtyard – to let in light and provide space for a well and room for farm animals

as far north as 32 degrees. The mainland of China is bypassed by the current and so has little or no reef developments, but fringing reefs do grow on the southern tip of the tropical island of Hainan and around other islands in the South China Sea, where there are also hundreds of coral atolls. The closer the islands to the Southeast Asian waters, the more diverse are their corals.

Atolls are rings of coral with lagoons, which originally formed as fringing reefs around deep-sea volcanoes. As a volcano gradually subsided, the coral would have continued to grow, keeping near the surface, until all that was left was a ring of coral around a lagoon where the volcano had been. The South China Sea atolls and the fringing reefs are thought to support at least 2000 fish species and 130 coral species.

Each coral polyp animal can draw energy from the sun using single-celled algae – zooxanthellae – in their bodies, but they also feed on plankton, catching it with harpoon-like nematocysts. The reefs they form are highly complex ecosystems – the tropical rainforests of the marine world – with the photosynthesizing corals providing food and shelter for others. Though typhoons frequently break up and kill the coral, leaving fragments of their hard internal structures scattered over the reef, these white skeletons, like fallen trees in a rainforest, provide platforms for new growth. Young coral polyps settle on them, building new coral gardens on the bones of the past.

The kaleidoscope colours of live corals are produced by their internal algae and enhanced by colourful fish that dart among their branches. Life for the fish depends on the microscopic phytoplankton that drift up from the deeper water in and around the reef. On them feed zooplankton – from copepods to comb jellies and planktonic fish larvae. The action of the monsoon supplies the sea with larvae from many different sources, and so it's possible that a fish living on one reef originally came from one 1000km (620 miles) away. The combination of these shallow-water coral reefs and the deep-water, plankton-rich zones in the South China Sea are what have, until now, ensured a constant supply of fish for the commercial fisheries.

Where the plankton-rich currents flow over the atolls, filter-feeders abound – sponges, worms, feather stars and molluscs such as the giant clam, hard and soft corals and crustaceans. The biggest filter-feeders of all in terms of the volume they consume are the manta rays and whale sharks. But some of the larger predators, in particular the sharks, are now rare – a result of China's increasing consumption of shark-fin soup. The demand for seafood has also devastated the coastal reefs, where

a large number of shallow reefs, such as those in the Xisha (Paracel Islands) and Dongsha Islands, have been bombed and bleached. Dynamite is used to stun fish, which then float to the surface, and cyanide has been used to collect fish for the aquarium trade in the South China Sea.

The problems are illustrated by the threats to Hainan's fringing reefs. Originally inhabited by aboriginal Austronesian peoples, Hainan was seen as remote and treacherous, a perfect place to banish exiles to. Its isolation preserved its natural heritage until recent years – a heritage that has made Hainan the second hottest tourist destination in China. Its major attraction is Sanya, 'the Hawaii of the Orient', with new four-star hotels, pools, palm trees and Yalong Bay, offering 7.5km (4.7 miles) of white sandy beach. This and locations such as the nearby Sanya National Coral Reef Nature Reserve attracts up to 1.5 million visitors a year.

The Sanya reserve, established in 1990, is one of only three coral reserves in China, largely saved from the fate of the surrounding area, where the reef fish and corals have been destroyed by a mixture of siltation and overfishing. But tourism itself is a potential threat. While on the one hand, it has resulted in management of

ABOVE
A female Hainan eastern black-crested gibbon (males are black), one of only a few dozen left, hanging on in the forest of Bawangling Nature Reserve. This is one of the most endangered populations of primates in the world.

the reserve and increased awareness of coral reefs, the number of people visiting the reefs – and demanding seafood and seashell trinkets – is astounding. One organization alone sends hundreds of first-time divers a day to the reef.

Hainan also has a green heartland, home to some of the world's rarest creatures, including the Hainan eastern black-crested gibbon. Conservation projects involving local people are helping to save the remaining forest and protect the animals from poaching. There is also an ambitious plan to plant thousands of hectares of mangrove trees along China's shores, from Hainan to Fujian, which will provide natural buffers from the typhoons and tsunamis of the future, as well as nurseries for fish and roosting sites for migrating birds.

coastal development

The Bo Hai Gulf has been called the 'fish, salt and oil storehouse'. Its fertility comes from the nutrients pouring into the sea from 40 rivers. And the fact that the average depth of the gulf is just 18 metres (59 feet) means that nutrients are easily stirred up by surface winds, creating a rich soup for fish and filter-feeders. Both here and in the Yellow Sea and East China Sea, a copious supply of silt encourages the growth of phytoplankton, which form the base of the food chain. They depend on specific conditions for growth and so can be the first indicator of pollution or temperature changes. Phytoplankton blooms caused by high nutrient levels – sewage or fertilizer run-off, for example – can change the colour of the sea to such a degree that the blooms can be seen from space.

Red tides are algal blooms that turn the water red or brown. Some of these are produced by toxic algae, which can contaminate shellfish, kill fish and poison humans. China experienced 83 toxic red-algal tides between 1990 and 2004, and the fact that this phenomenon is continuing is a sign of a worrying change in the sea's ecology.

Few wild fish or crabs survive in the Bo Hai Gulf in significant numbers because of pollution and overfishing. But there is one 'success' story. Every July, 700,000 fishermen prepare their boats for a special fishing season. They are after jellyfish, which overwinter as juveniles buried in the sediment and move into the water column in the spring to start growing. The fishermen can haul in enough jellyfish in a week to keep their families in money for many months.

At a signal from the government, the boats sail out to find the best place to lay their nets and haul in the drifting jellyfish. They are looking for two particular species, which don't sting and are highly sought after as appetisers. The tentacles are the delicacy and are separated from the mantle with a swift turn of the wrist and then layered with salt. Some are offloaded onto the sandy beaches around Yinkou port, where hundreds of mules wait to be loaded up with jellyfish before struggling up the shore to the sound of whips cracking the air. The jellyfish are rinsed and sliced and then usually eaten raw with a touch of chilli, coriander and vinegar – a dish with the sweet taste of oyster and the crunchy texture of an iceberg lettuce.

The likelihood is that the jellyfish are plentiful because of the overfishing of plankton-eating fish such as herring – competitors for the jellyfish's food – together with warmer waters and the jellyfish's ability to thrive in polluted waters.

TROUBLE IN THE YANGTZE

the toxic water comes from the output of factories, farms and cities that pour into the river, making the river the biggest source of marine pollution in the Pacific

The catchment of the Yangtze – the longest river in Asia – covers a fifth of the land area in China and accounts for 40 per cent of the fresh water. Its delta has the most fertile soil in China. Rice is the main crop, but the river also provides more than 70 per cent of fishery production. It is also home to at least 350 fish species, of which 112 are found nowhere else.

One of the biggest fears of fish farmers in this region is that toxic water will seep into their man-made lagoons. The toxic water comes from the output of factories, farms and cities that pour into the river, making the river the biggest source of marine pollution in the Pacific. These toxic substances threaten both the wild fish and the animals that feed on them, including the endangered Chinese sturgeon, the endangered finless porpoise and the baiji, or Chinese river dolphin, which in December 2006 was declared functionally extinct after an extensive search of the river. The baiji joins other animals such as dugongs, otters and saltwater crocodiles, which now appear to have all but vanished from China's shores.

Great strides are being made to improve the quality of water in the Yangtze, but the task is huge. According to WWF, the river receives 42 per cent of the country's sewage discharge and 45 per cent of its total industrial discharge. Shipping discharges add to the pollution. Loss of the floodplain to agriculture has also reduced the basin's ability to detoxify pollutants.

The government is now taking steps to develop a management plan, involving restoration of the floodplain wetlands and more sustainable agricultural practices.

the urban transformation

In 2010, for the first time in history, humans will be predominantly urban, and China is leading the trend. Coastal provinces currently hold 700 million or so of China's 1.3-billion-plus people. In the next 25 years, it's thought that 345 million more will migrate from the inner countryside to the thriving cities – the largest-ever human migration. This doesn't bode well for the coast's remaining natural areas, though with help, at least some wildlife will survive. Take Hong Kong as an example.

The island is one of the most populated places in the world, but 40 per cent of its land is conserved as parks and special areas. Mai Po Marshes Nature Reserve is a plot of wetland on the edge of Inner Deep Bay, jutting up against the high-rises of Shenzhen city. Mai Po was originally used by gei-wai (shrimp-pond) operators. When the shrimp-harvesting season ended in early winter, they drained their gei wai to get the by-catch of fish. The exposed mud would then provide a feeding ground for hundreds of fish-eating birds, particularly herons, egrets and the endangered black-faced spoonbills. WWF has continued to drain one gei wai every two weeks from November to March to provide feeding grounds for migrating waterbirds. The reserve now attracts 25 per cent of the world population of black-faced spoonbills – on the brink of extinction until reserves were set up along China's coast to provide stop-off points.

Planks on floating barrels take you over the mangrove mud to the hides that overlook the estuary. If you visit in winter and early spring, you will see an aerial ballet of up to 68,000 birds rising and falling as they jostle for the best feeding positions, moving with the tide. The birds – from black-winged stilts, stints, sandpipers and knots to godwits and curlews – have flown down the coast of China from as far as Siberia.

ABOVE
Thousands of waders flying up to follow the tide at the Mai Po Marshes Nature Reserve. In the background is the city of Shenzhen – one of the fastest-growing cities in China, bordering Hong Kong.

Only the chugging boats and high-rise buildings are reminders of the city next door.

The Pearl River delta to the west of Hong Kong is a major manufacturing base for electronic products. It is polluted with sewage and industrial waste, and much of the area is frequently covered with brown smog. This pollution may have affected the birds at Mai Po by reducing the numbers of crabs, shrimps and fish on the mudflats, and it is definitely affecting one of the region's most famous animals – the Chinese white dolphin, or Indo-Pacific humpback dolphin – living around Lantau Island west of Hong Kong, thought to number only about 200 in the whole estuary.

Usually seen in groups of three to four, the dolphins are born dark grey and become spotted as adolescents and creamy white or dark pink as adults. The pink colour is probably due to the fact that, when grey pigment is lost as the animals mature, the capillaries show through the skin (as when a human blushes). The territory of the Pearl River populations includes busy shipping channels with 70 or so vessels an hour passing through, and the dolphins must endure the deafening sounds of blasting, dredging, sonar and shipping, including high-speed ferries. Recent research has suggested that they may have adapted their communication to compensate for noise pollution, packing more information into shorter calls.

The largest coastal city by far is Shanghai, the centre of Chinese economic development. In ten years, Shanghai has built the same number of skyscrapers as in the whole of New York. Last year, half of the concrete used in construction around the world was poured into China's cities and roads, and the demand for new apartments, office blocks and skyscrapers is so intense that work continues round the clock. The Huangpu River running through Shanghai is filled with barges loaded with cement for the building sites. Shanghai is not shy in proclaiming its success by

BELOW
A Chinese white dolphin and her offspring near Lantau Island off Hong Kong. This population is threatened by boat traffic and pollution in one of the world's busiest harbours.

THE WHISTLING HUNTER

Chongming Island in the mouth of the Yangtze is the largest alluvial island of any river mouth in China. Dongtan Nature Reserve at its eastern end is a spring and autumn wetland stop-off for 2–3 million migrating birds. But Shanghai's new wealth means that the rare and exotic are now in demand for the banquet tables of its richest citizens, and the biggest threat to the migrants is now human predation.

Mr Jin has been a hunter for many years. Using a handmade bamboo whistle, a couple of decoy birds and a flip-net, he catches waders. He sets his net on the mud, perches on a stool under a camouflaged parasol and waits patiently. He scans the open skies for wading birds, and once he spots them, mimics their songs with whistles. Mr Jin's remarkable repertoire includes the calls of up to 30 species. The birds fly towards the decoys, and with a swift pull of the line, he traps them under the net.

But the birds are not destined for the pot. Mr Jin now works with government-backed conservation teams, catching birds for tagging. The birds are weighed, measured and then released to continue their journeys.

> his remarkable repertoire includes the calls of up to 30 species. The birds fly towards the decoys, and with a swift pull of the line, Mr Jin traps them under the net

revelling in its night illuminations. Though stunning, they are also a reminder of the many coal power stations being built every year to fuel such excesses.

Recently, though, China has taken its first steps on a long journey towards more ecologically sound development. On the world's largest alluvial island, Chongming, just 25km (15 miles) from Shanghai, the world's first eco-city is being built. Dongtan has been conceived as a series of towns connected by cycle routes and public-transport corridors. The city will be largely powered by renewable energy – sun, wind and biomass. One reason for the decision to create this innovative city is the existence of a huge wetland area on the eastern end of the island – a reserve for migrating wading birds, where more than 250 species have been observed.

China is in the middle of a cataclysmic change as significant as that of Britain's industrial revolution two centuries ago, but it is attempting to refashion itself in less than 50 years. The result is pollution and the loss of architectural heritage and natural wealth. But more than any other country in the world, China has the people and the power to make the changes necessary to reduce the impact of its consumption on the environment – if the will is there to do so.

in the next 25 years, it's thought that 345 million
more will migrate from the inner countryside to
the thriving cities

the
gazetteer

WHAT FOLLOWS is a selection of places of natural, landscape or cultural interest, some famous, some not so, where we filmed during the making of the *Wild China* BBC series or which we recommend visiting. It can only be a taster, since China covers an area about the size of the USA. It has 23 World Heritage Sites, more than 1400 nature reserves, more caves than any other country, the highest mountain on Earth and the world's largest man-made structure – the list is endless. In short, whatever you are after, this land of superlatives has it.

The People's Republic of China is also home to a fifth of the world's population and has some of the most polluted cities on the planet. It is divided into 22 provinces and 5 autonomous regions in which ethnic minorities other than Han Chinese predominate (there are 56 official ethnic groups). Unless you have unlimited time and budget, it's sensible to explore one or two regions rather than ticking off all the highlights in one trip, this will allow you to immerse yourself in China's 5000-year-old culture. (Bear in mind that some areas are still effectively closed to foreigners.)

The biggest barrier you are likely to encounter is language. Picking up the subtleties in a short time is almost impossible, and so it's advisable to have important information and useful phrases such as your destination or hotel name

The 1200-year-old walls of the imperial city of Pingyao – one of China's many World Heritage Sites.

written down in Chinese. That said, take a phrase book, as even a simple hello or a thank you goes a long way. Within Chinese place names, look out for the following suffixes: Shan – mountain range; Jiang – river; Hu or Tso – lake; and La – pass.

You'll find organized excursions at most tourist destinations, but independent travel within China is surprisingly easy. Nearly everywhere is accessible by roads that are generally in good condition – a mixed blessing, enabling people to travel too fast with few driving skills. The bus and rail network is extensive and fairly cheap but varies in quality. Most large cities have airports, and internal air travel is relatively straightforward and avoids long, monotonous bus or car journeys, often involving travel at night.

Accommodation ranges from five-star palaces in Shanghai to filthy, flea-ridden truck-stops. As a rule, three stars and above is OK, but be prepared for some very average hotel experiences and even more average sanitation. When eating, be aware of hygiene. Fantastic hospitality and rewarding experiences can be had by staying with local people. In remote regions, camping can be the best and only option, but check that it is allowed where you plan to stay. The Chinese have an obsession with food, and the cuisine is some of the finest and most diverse. In rural areas and back alleys, be open-minded about what appears on your plate – and be prepared for weird delicacies.

Many of the wildlife highlights listed are in nature reserves or off the beaten track, and so access varies, and permissions may be needed.

Seeing wild Chinese wildlife is always difficult, and what wildlife you may see is often managed to cater for the rapidly expanding internal tourist industry. Chinese tourists travel en masse, especially on national holidays, and so keep this in mind when planning. You'll certainly have to work hard to find peace and tranquillity, but the experience of being there and allowing time to soak up the culture and atmosphere of a region will not only leave you dining out on memories for years but will also give you an insight into arguably the most fascinating and rapidly changing country on the planet.

1 the heartland

Think of central China as a sea of people interspersed with islands of cultural wonder, and you won't be far wrong. The fertile soils of the heartland spawned the birth of agriculture 8000 years ago and possibly human civilization itself and now support hundreds of millions of people, most of them Han, the largest ethnic group in the world. Though some of the world's most polluted land can also be found here, islands of wilderness still exist. Travel between these can be monotonous, but airports are plentiful and roads good.

WILD WONDERS

crested ibis

Thanks to Chinese conservation efforts, a healthy population of several hundred of these extremely rare birds almost guarantees a sighting. Fly to Hangzhong and drive an hour to Yangxian county town. The local breeding centre, 3km (1.9 miles) north of Yangxian, will be able to recommend the best viewing spots. The birds nest in April and May, and mid-May is a good time to see young being fed at the nest. In autumn, look for the roosting tree just outside Caoba village, close to the breeding centre, where about 60 birds roost each night.

giant panda

There is no better place to see giant pandas than in the famed Wolong Giant Panda Breeding and Research Centre (open daily) in southwest Sichuan province. At the time of writing, 18 cubs were cavorting in one enclosure – a vision to tug the heartstrings of even the hardest. The four-hour drive from Chengdu over the Balang Shan range is an experience in itself but can be hairy, with roadworks and landslides. There are hotels on site

Babies on show to visitors at the Wolong Giant Panda Breeding and Research Centre.

and 7km (4 miles) away. While in Wolong Reserve, walk along the Min River – a haven for birds. You can buy trekking maps, but trails can be difficult to follow.

golden monkey

Zhouzhi Nature Reserve lies on the northern slopes of the Qin Ling Mountains. Here you can sit and watch the world's only fully habituated troop of golden monkeys. Their shaggy golden coats, blue faces and range of expressions make them among the most endearing of monkeys. They occupy family units of about ten. If you're lucky, you'll be in Zhouzhi when these groups congregate together in a super-troop of more

than a hundred. Zhouzhi is a four-hour drive from Xi'an. When the tarmac runs out, be prepared for a one-and-a-half-hour, cliff-hugging, bone-shaking ride on three-wheeled motorbikes. Don't arrive late, as headlight use is rare. Arrange the trip in advance, and stay overnight at the basic but clean accommodation. Tours for the viewing area depart before 10am and involve a one-hour walk up steps that never seem to end in temperatures that, in June (probably the best time to go), can reach 30°C (86°F), and so take water. Though the golden monkeys are herded to the viewing area by local farmers, they soon settle down in the trees above you and

MONGOLIA

G O B I

Hohhot ○

Jingshanling ○
○ Simatai
BEIJING ●
Tianjin ○

Bo Hai
Gulf

Ordos
Desert

Yellow River (Huang He)

SHANXI

○ Shijiazhuang

○ Taiyuan

SHANDONG
Yellow River
△ Jinan
△ Tai Shan
○ Tai'an

Qinghai
Hu

○ Pingyao

○ Lanzhou

Hukou
Waterfall ○ Jixian

GANSU

SHAANXI

Shaolin ○
Temple ○ Dengfeng

○ Zhengzhou

Wei
○ Xi'an
Zhouzhi NR ○
Mt Taibai △ Qinling
Mountains

HENAN

Juizhaigou NR ○
Jiu Huang ✈
Airport

Hanzhong ○ ○ Yangxian

ANHUI

Yangtze (Chang Jiang)

Balang Shan △
Wolong NR ○

SICHUAN

Three Gorges
Dam
|

HUBEI

Wuhan ○

○ Chengdu

CHONGQING

Leshan
Emei Shan △ ○

Chongqing ○

HUNAN

Poyang
Hu

0   100   200 kilometres

0   100   200 miles

occasionally come down to drink relatively close to you. Infants and juveniles spend hours playing together in arboreal crèches. You'll have several hours of quality time with them before they amble off into the broadleaf woodland. Listen out for their unique calls, made without opening their mouths. The whole experience is magical.

golden takin

While the pandas occupy the middle ranges of the Qinling

One of the gems of the Qinling Mountains – the golden takin, found only in China.

Mountains, the peaks are reserved for the less well known but equally spectacular golden takins. These huge animals are best seen in mid-June, when herds of up to a hundred congregate in grassy areas for rutting. They emerge from the dense bamboo forests to graze in the early morning and late evening. Males are particularly impressive, but they may attack if surprised in the long bamboo, and so employ a qualified guide. The walk up to the top of the Qinling Mountains will take five to six hours, depending on your chosen location, and requires sturdy footwear. Specialist tours are available.

Though local farmers will act as porters, this is certainly no day trip. Allow at least three days for the experience, which is well worth the time. You'll be camping in the wilds at more than 3000 metres (9840 feet) and trekking along stunning mountain ridges. Avoid July, when the rains start, making the mountain-tops treacherous.

LANDSCAPE WONDERS

Great Wall
The wall is one of the world's great wonders and a UNESCO World Heritage Site, and so it is besieged by tourists. It is actually a series of walls, and the quietest and most 'natural' section – also the steepest – is only two hours' drive from Beijing. Visit Simatai in winter, and you'll be almost on your own. Stay and eat at the simple accommodation at the centre – a 15-minute walk from the base of the wall. Leave Simatai at dawn for sunrise on the wall, and arrange to be collected at Jingshanling in the afternoon. This 10km (6-mile) walk will take five hours, and so bring food and water, and don't attempt it without a basic level of fitness or if your balance isn't too good.

To sleep on the wall requires permission, and so consider booking the trip through a travel agent, but it will add that something extra – especially at full moon – to what will be an experience of a lifetime. Though winter is the quietest time, summer is the most beautiful, especially at dawn, when the wall emerges through a sea of clouds.

Hukou Waterfall
The best place to see the Yellow River is at Hukou Waterfall on the border of Shaanxi and Shanxi provinces, where the sluggish, wide river narrows dramatically. The waterfall is not that high, but the volume, colour and roar of water in summer (July to September), when river levels are highest, is impressive. Or go for the midwinter spectacle of a frozen waterfall. The falls are a six-hour drive from Xi'an, which you can fly to. On-site accommodation is limited, but you can stay in Jixian, 25km (16 miles) east.

Qinling Mountains
The mountains are shrouded in some of the most biologically diverse temperate forests in the world, which offer a chance of seeing wild pandas and a wealth of other animals and plants, including more than 200 species of birds. The main peak, Mt Taibai, rises 3767 metres (12,358 feet) above sea level. Walking in the mountains – some of which are very steep – is physically challenging and requires waterproofs, first-aid gear and stout walking boots. Avoid July and August, when rain is most common and the pandas have migrated to their summer ranges above 2000 metres (6560 feet). The mountains are a four-hour drive from Xi'an, which you can fly to. There are a number of reserves, and tour operators can arrange wildlife tours and basic accommodation.

Jiuzhaigou Nature Reserve
This is a group of three picture-postcard alpine valleys, about 400km (248 miles) north of Chengdu in Sichuan province. Jiuzhaigou is famous for its crystal-clear blue, green and turquoise-coloured lakes, complete with underwater fossilized tree trunks. The area is cloaked in broadleaf forests and topped with snow-capped mountains. Take a bus to view some of the sights or lose yourself on a tranquil walk. Though beautiful all year round, the best time to go is autumn – the second or third week of October – when the glowing autumn colours are reflected in the mirror-like lakes. There are lots of

Jiuzhaigou in early autumn, the best time to go. There are beautiful walks and tranquil lakes.

hotels in Jiuzhaigou, but book in advance, as the area is very popular with Chinese tourists. Fly to Chengdu and take a 12-hour bus journey on the beautiful, sinuous road that follows the Min River. Alternatively, fly 40 minutes from Chengdu to Jiu Huang airport and drive the remaining 85km (53 miles) – one and a half hours – to the entrance to Jiuzhaigou.

Emei Shan

This is one of the four sacred Buddhist mountains, dotted with temples, pilgrims and the famous Emei Shan macaques. When passing through the 'joking monkey' section of the walk to the summit, watch out for the overfriendly simians, who like to help themselves to your lunch if you're not looking. As with any mountain, weather on Emei Shan

can change quickly and dramatically. Make sure you bring clothing for all seasons and solid footwear. The best time to climb the mountain is spring or autumn, but you can hire iron crampons if a winter wonderland is your goal.

Fly to Chengdu and drive 80 miles (129km) south to Emei town. From here, it's another 4 miles (6km) to the base of the mountain. You'll need at least a day (and possibly several) to climb the 3100-metre (10,168-foot), vertigo-inducing, golden summit peaks, and a day to clamber back down to safety. For those who lack the energy for this climb, a minibus ride from Baoguo and then a cable car will transport you to the summit to marvel at the awe-inspiring views. There are many temples en route that offer food and a bed while you plan your next day's

adventure, but don't expect heating and other such luxuries. Just below the summit, you may have the chance to see 'Buddha's glory': as sunlight catches the clouds around you, your shadow will take on a rainbow halo.

CULTURAL WONDERS

Beijing's parks

Early morning visits will avoid the tourists that throng many of Beijing's attractions and will also give you an insight into traditional life in China. The parks act as focal points for Beijingers and are well worth a visit, particularly at dawn and dusk on weekends. Travel by taxi – relatively cheap, but the driver won't speak English, and so take a written form of your destination and accommodation. The following are just a few park highlights.

The entrance to a Beijing park haven.

Strange sounds echo round the **Temple of Heaven Park** at sunrise as people go through their private tai chi routines to exercise and help maintain harmony throughout the day. In winter, between November and March, look for the long-eared owls in the ancient cypress trees just east of the Altar of Heaven.

Autumn is a great time for a quiet stroll among the gingko trees in **Ditan Park**. Watch the men writing poetry on the flagstones. Visit the vibrant, noisy fair that celebrates the Chinese new year in January/February (the date differs each year, depending in the date of new year).

Central to **Jingshan Park** is a 48-metre (157-foot) artificial hill made from the rock and soil excavated from the Forbidden City's

moats, which gives unparalleled views over the Forbidden City and the rest of Beijing. Sit up here to watch the sun set on a Sunday evening, accompanied by the sound of hundreds of people singing traditional Chinese folk songs beneath the ancient gnarled cypress trees in the park below. As dusk falls, watch the bats silently stream out of the roof of the east gate of the park and into the Beijing night.

Beijing hutongs

Hutongs are traditional housing communities – low, redbrick houses around a central courtyard – that are fast disappearing in Beijing as they are demolished to make way for high-rise apartments. Tours can be arranged from most hotels, but going on your own can be more rewarding. Visit on a summer evening, when locals sit outside playing card games and xiang qi (Chinese chess). Crickets hang in wicker cages above them. The insects apparently enhance the ambience, and a good singer is prized. You are likely to be beckoned over and given a stool and a beer – and this is by poor people who don't know you. Go with a Chinese speaker to gain an insight into their way of life. After about 10pm, you may notice another visitor – the Siberian weasel, revered by some as a spirit.

Pingyao

This imperial city in Shanxi province is a UNESCO World Heritage Site. Visit some of the many museums for an overview of local Chinese history to mull over while you walk round the city's 1200-year-old walls.

The new-year dragon dance at Pingyao, best watched from the city walls.

The amassed replica army of the first Emperor of Qin. The terracotta figures are life-sized.

Or hire a bicycle from the main street. The train takes 10 hours from Beijing, but you can fly to Shanxi's capital, Taiyuan, and drive from there. Accommodation is plentiful.

At the Chinese new year, watch the dragon dances from the city's 20-metre-high (65-foot) walls, while fireworks light the sky around you and the city hums and crackles in celebration below.

terracotta warriors

This is one of the world's most famous archaeological discoveries (made in 1974 by local farmers) – a 2000-year-old terracotta army of 8099 life-sized, heavily armoured warriors, with individual clothing and expressions, together with their horses. It guards the pyramid tomb of the first Emperor of Qin (Qin Shi Huangdi), interred in 210-209 BC, and yet to be excavated. Take a bus there from Xi'an, 35km (22 miles) away, or hire a tour. Xian has plenty of accommodation, and there are many fascinating sites to visit in the area. Go in winter to avoid some of the crowds.

Shaolin Temple at Song Shan

A visit to this sacred mountain offers a range of ancient Buddhist temples and the pagoda forest tombs of 240 eminent monks and abbots. The area is famous as the birthplace of kung fu, a martial art developed about 1400 years ago by Shaolin monks to protect themselves and the temple. There are daily kung fu shows and even classes at nearby Dengfeng. The mountain also offers awe-inspiring walkways that literally hang on the edge of Song Shan's sheer cliffs. The climb is arduous, but you can take the cable car that starts not far from the Shaolin Pagoda Forest. There is also a chairlift in summer. Song Shan is an uneventful ten-hour drive from Beijing, or fly to Zhengzhou and drive the remaining hour to Dengfeng – the ugly but functional town which is your base. You'll be woken at 6am by the practice sessions of kung fu schools in the town, which are a slightly scary spectacle in themselves. The area is a year-round Chinese tourist mecca. So expect hordes of noisy groups. It is perhaps most beautiful in spring (May) and autumn (September-October). Avoid public holidays and weekends.

2 north of the wall

Northern China is about as varied a region as is possible – from the frozen forests of Heilongjiang to the hubbub of Kashgar and the vastness of the Gobi. Don't expect to see much wildlife, but the people and landscapes more than make up for this. Xinjiang Uyghur autonomous region has the highest density of attractions, and the Silk Road and the eight international borders make it a cultural melting pot. You'll have to travel long distances and endure extremes of temperatures, but it is worth the visit for anyone with a sense of adventure.

WILD WONDERS

giant gerbil

These are relatively common throughout northwest Xinjiang, but one of the most convenient places to watch giant gerbils is on the outskirts of Turpan, beside the carpark of the botanic gardens (containing 400 or so desert plants, and worth a visit in itself).

The ground is covered with their holes, and you should be able to watch them busily gathering food

A wild Przewalski's stallion seeking out a receptive mare in Kalamaili Nature Reserve.

The 'carpark' giant gerbil.

A kiang, or wild ass – one of a number of mammals you may see in the Kalamaili reserve.

throughout the day. At night, use your car headlights to watch jerboas foraging, hopping around on their long hind legs.

Przewalski's wild horse

All Przewalski's wild horses alive today are directly related to a handful of animals captured at the turn of the twentieth century and bred in European zoos. If your goal is to see horses in the wild, head to the Kalamaili Nature Reserve – 400km (248 miles) along the highway that runs north from Urumqi round the east of the

Junggar Basin. Keep an eye out for the Przewalski's horses about 30km (19 miles) from Jiakuerte.

There are now more than 40 horses in the reserve, and a new batch of foals each summer (avoid visiting in winter, as heavy snowfalls and temperatures of -40°C are not unknown, and the horses are kept in a big enclosure).

While you are in Kalamaili reserve, keep an eye out for goitered gazelles, wild asses and a wealth of birds of prey including golden eagles and saker falcons.

whooper swans at Bayanbulak

In April, several thousand whooper swans arrive at Swan Lake, a large wetland within the Bayanbulak (Swan Lake) Nature Reserve. They breed around the lake, moult in July and August and then congregate in large colonies in September while waiting for the young to be strong enough to migrate. Most of the Chinese tourist groups have gone by September. The reserve is a 40-minute car ride from the nearest basic hotel. Or stay in Mongolian ghers for the night and ride out

A whooper swan by one of the Bayanbulak lakes.

Sunset over the Kaidu River in the vast area of Bayanbulak, a prairie grassland with lakes and rivers including the wetland Swan Lake Nature Reserve.

around the stunning Kaidu River on horses (available for hire). This is a picture-postcard landscape and also offers the opportunity to spot up to 120 bird species. Bring warm clothes, because temperatures can drop to below freezing, even in summer. The reserve is 270km (167 miles) northwest of Korla, and you can organize a trip from there or from Ürümqi, though the drive will take about ten hours.

Tianshan Wild Animal Park

An alternative to travelling huge distances looking for wildlife is to visit the new Tianshan Wild Animal Park in Xinjiang – the largest zoo park in China and just 25km (15 miles) from Urumqi on the south slope of the Bogda Mountains. It is divided into desert, mountain and grassland biomes, and houses quite a number of native species, including wild horses, wild asses, goitered gazelles, wild camels, ibex, blue sheep and argali, all in enclosures that replicate their natural environments.

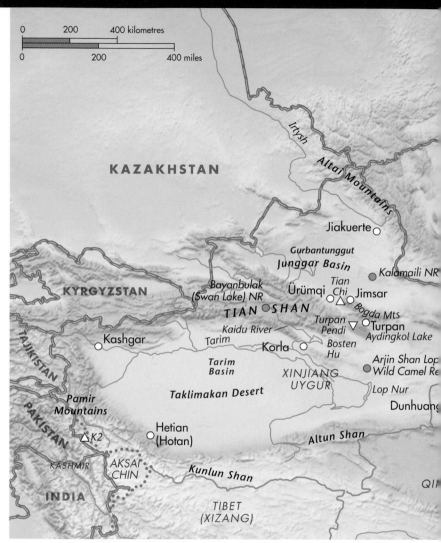

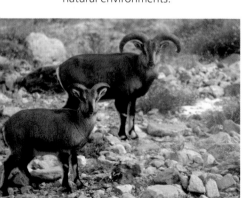

Mountain blue sheep – snow leopard prey.

Manchurian, or Siberian, tiger

This endangered cat is the largest in the world. Though it's almost impossible to see one in the wild, many captive-bred Siberian tigers are housed at the Hengdaohezi Felid Animals Raising and Breeding Centre, in the Songbei development zone, 15km (9 miles) outside Harbin in Heilongjiang province. If you decide to visit, be prepared to see many overweight captive-bred tigers as you are driven around the park in somewhat rickety vehicles. Be aware that Chinese visitors sometimes pay for live animals – including cows – to be thrown to the tigers, and that many tiger farms want to open up the trade in tiger bones. Finally, go in the knowledge that captive-bred tigers are highly unlikely to be successfully released to the wild.

Tian Chi in the crater of Mt Baitou volcano.

LANDSCAPE WONDERS

Changbai Shan Nature Reserve

This fairytale mountain range is situated in the largest national park in Jilin province. Its focal point, straddling the Chinese/North Korean border, is Tian Chi (Heaven Lake), a high-altitude caldera (not to be confused with Tian Chi in Xinjiang province – see p220) filled with crystal-clear water. On the way to the lake, you'll pass hot springs and stunning waterfalls. There is even a place where you can soak in steaming volcanic water.

Chinese and Korean tourists occasionally swamp the main meeting areas at the lake and the base of the waterfall that drains from it. Your best option then is to head off the beaten track, though make sure you know where the border is and where you can and cannot go. Visit between June and September. In June and early July, the meadows on the west slope (75km – 47 miles – from the lake) bloom with alpine flowers. Also, as with any mountain environment, the weather can change in a flash. So take food, water, sunscreen and medical supplies, and think about hiking in a group. To get to Changbai, fly to Yanji and then drive or take a train to Baihe. You can rent a car and driver to get to the lake, or take a bus from outside the train station. Base yourself in Baihe or book the more expensive accommodation in the reserve.

Jiayuguan

Throughout the Ming dynasty, Jiayuguan – known as 'the mouth of China' – marked the western edge of the Han empire. Thought by many to be the end of the great wall (the wall did stretch further west, but most of this has succumbed to the shifting sands of the Gobi), the town was the last major oasis for merchants and travellers wishing to travel to Xinjiang and beyond. Cycle out to the impressive fort 6.5km (4 miles) to the west, which offers possibly the iconic image of western China. Climb to the top of the fort and walk along its battlements to enjoy the impressive scenery of the Qilian Mountains 129km (80 miles) away. You can fly direct to Jiayuguan airport, 13km (8 miles) north of the city. Otherwise, it's a six-hour

journey from Dunhuang or an overnight train from Lanzhou. There is a range of accommodation in Jiayuguan city and organized trips to the fort or the nearby glacier in the Qilian Mountains. If you visit the glacier, come prepared for cold weather and altitude, because you will be hiking at over 4300 metres (14,105 feet) and will feel the effects of oxygen deprivation.

yardangs

Centuries of howling winds and rain have eroded the Taklimakan Desert and Lop Nur and blown much of it east towards Beijing and beyond. What are left are yardangs – strangely symmetrical shapes formed as sand scours hillocks of harder substrate into long, streamlined pillars. Wind blowing through the yardangs can create strange howling noises – hence the name devil city. The best place to see them is the Dunhuang Yardang National Geological Park, just 180km (112 miles) from Dunhuang. You can stay in Dunhuang and take a day tour to the desert. Winds are strongest in March, but tours may be cancelled at this time of year, and so come in the summer (May to September).

Tian Chi (Heaven Lake)

Try camping with Kazakh nomads and their camels on the shores of

Heaven Lake surrounded by the peaks of Tian Shan. You can camp by the shore in Kazakh yurts.

A Kazakh family breakfasts on warm mare's milk before packing up to move to winter pastures.

Tian Chi in Xinjiang province. You can stay in yurts surrounded by beautiful coniferous forests and the lofty peaks of the Tian Shan range. The area is great for trekking and camping, but bring cold-weather clothes, food and water.

Visit in early September, when the nomads move down from their summer pastures to their low-altitude wintering regions. If you plan it right, you can time your stay with their departure and walk with them for some of the way. The sight of camel trains loaded with the nomads' worldly possessions will make you a privileged spectator in a practice that has remained unchanged for centuries. Tian Chi is 120km (76 miles) from Ürümqi, a three-hour journey and an hour's hike up the hill – or a cable car, if you're feeling less active.

CULTURAL WONDERS

Harbin ice festival

If you want to feel the full force of a Chinese winter while experiencing a unique spectacle, head to Harbin – the capital of Heilongjiang province. The architecture provides a stark reminder of Russia's former presence, but in January or February, depending on the Chinese new year, the city is transformed. The whole of Zhaolin Park is filled with strange animals and buildings carved out of ice and illuminated into a multicoloured extravaganza.

Prepare to be amazed but also to be frozen – temperatures can drop to below -30°C (-22°F), and so arctic clothing is a must. Some locals get a kick out of swimming in the frozen river, entering the water through ice holes.

Chagan ice-fishing festival

In mid-December, the frozen Chagan Lake in Jilin province launches the ice-fishing season. First comes the 'waking the lake' ceremony, involving Mongolian dancing and chanting, and then hundreds of people walk onto the lake to select their fishing areas. Holes are cut into the thick ice, and 500-metre (1640-foot) nets are inserted. Next day, the crowds gather again as nets are hauled in by a combination of horses and tractors, each net yielding up to 5 tons of fish. The season runs for about a month. Get here for the first day and join in the celebrations, which can go on for days. You can fly to Changchun, the capital of Jilin, and drive the 150km (93 miles) or so to the lake. Tour operators in Songyuan will have the fishing-season dates.

Nadam horse festival

Throughout history, Mongolian warriors have been famed for their expertise in the saddle. Using an army of cavalry, the mighty Genghis Khan created one of the largest empires in history. The festival of Nadam (Mongolian for competition) – a celebration of Mongolian culture – occurs on grasslands throughout the province in July or August. Spend a day or two camping in a forest of ghers while watching the archery wrestling and long-distance horse races – some riders are as young as five. You can also ride across

Father and son discuss tactics before the Nadam horse race. The festival of Nadam occurs throughout Inner Mongolia in July or August.

the Mongolian steppes in tours organized from Hohhot, capital of Inner Mongolia. Nadam is a movable feast, but travel agents and hotels can point you in the right direction at the right time.

Mogao Buddhist caves

These desert caves, hewn out of the cliff face, are 24km (15 miles) southeast of Dunhuang – an oasis town on the edge of former Chinese Turkestan in western Gansu province. They were created and adorned with Buddhist imagery between AD 500 and AD 1000 and then mysteriously sealed in the eleventh century. They weren't rediscovered until 1900, when 50,000 parchments and murals were unveiled to the world. What makes the murals so interesting is that they reveal the evolution of Chinese art through the dynasties.

Of the 1000 or so caves, 600 are considered of interest, but only 30 are open to the public (some are considered too graphic for public viewing). Mind you, if you see 15 in a day, you'll do well.

You can plan a tour from hotels in Dunhuang or hire a bicycle for the day and pedal out to the caves.

grapes at Turpan

Nicknamed the oven, Turpan lies on the northern Silk Road in a depression descending to 154 metres (505 feet) below sea level. Temperatures can soar to 50°C (122°F) in summer, and yet the area is a centre for grape growing. The secret is the 2000-year-old underground irrigation system that brings water to the city from nearby glaciers in the Bogda Mountains. The karez (Uyghur for underground) systems are a shadow of what they were but still stretch for hundreds of kilometres, also providing a cool place for the locals to shelter in during the day. To savour more than 100 different grape varieties, head to Grape Valley 15km (9 miles) from the city. Or else just enjoy the Uyghur food and wine and amble or cycle through the city's vine-covered streets. Keep an eye out for red-tailed gerbils eating grapes from the drying sheds. Avoid midsummer and also winter (November to March), when temperatures can drop to -15°C (5°F). Possibly the best time to go is late August and September – harvest time. Turpan is a two-hour drive from Ürümqi; or catch a train to the nearest station 55km (34 miles) away and take a taxi or minibus to the city.

Kashgar

Kashgar is a Muslim trading Mecca, famous for its colourful bazaars. The market is open every day, but to experience the city's streets groaning with tens of thousands of traders and buyers, be out before dawn on a Sunday.

Haggle for carpets, knives, silk, jewellery and other miscellaneous wares with wily street traders, and sample the rich variety of cuisines that accompany the variety of cultures in this cosmopolitan town. But look after your camera, as pickpockets are active.

You can fly direct from Ürümqi or take the 24-hour train. Kashgar is perhaps the ultimate final destination of a trip to China. From here, laden with your wares, head out of China to Pakistan on the infamous Karakorum highway.

Kashgar market, full of traders of everything.

Karez underground water channels, thousands of years old and still bringing water to Turpan, where it can reach 50°C (122°F) in summer.

3 the tibetan plateau

Tibet. The word conjures up images of a magical Himalayan kingdom – a remote utopia. The truth, sadly, is somewhat different, and the people have endured a lot. That said, the Tibetan plateau remains a truly extraordinary place. It oozes cultural richness, the landscapes are magnificent – from grand vistas and multicoloured lakes to the great barrier itself – and the wildlife, though sparse, is unique and unforgettable.

Most of the plateau is above 4000 metres (13,120 feet). So expect wind and strong sun. Temperatures in the day can be warm, but at night they can plummet, and in winter they may go below -30°C (-22°F), with wind-chill on top. Lhasa is one of the highest cities in the world – but about as low as you are likely to get in Tibet. Oxygen levels are about 60 per cent of those at sea level, making even basic exercise difficult. Spend at least three days resting and sightseeing in Lhasa before moving to higher altitudes. The rule of thumb is to ascend 300 metres (1000 feet) a day and rest a day for every 1000 metres (3280 feet) ascent between sleeping points.

Travel is expensive and difficult, though the completion of the Qinghai-Tibet railway has opened up Tibet. If you can afford it, hire a 4x4 with guide and driver. Tour companies run trips – including cycling, hiking and climbing ones – from Lhasa (and Nepal).

You can travel independently, though research the legislation before you go. Permit regulations are prone to sudden change, and so check with the embassy or your travel operator just before you go, and be flexible.

On the plateau, you'll need to travel huge distances far from human habitation to see wildlife, as most species occupy a fraction of their former ranges in a fraction of their former numbers.

WILD WONDERS

argali sheep
This is the world's largest sheep, and the horns of the males are prized by international hunters, making the sheep wary of people. Possibly the best viewing is in the mountains of Gansu province. Fly to Dunhuang and drive south through the dunes for an hour to Aksai, at the base of the Tibetan Plateau.

Stock up on supplies, find a guide and drive to the heart of the Kharteng International Hunting Area – a day or so in a 4x4. You'll need to hike several days up and down the mountain peaks.

When walking along ridges, consider both wind direction and your silhouette. November is when the sheep come down from the peaks to mate and are less wary; there is also a chance of seeing males fighting. You may also see Kazakh nomads riding camels in snow as they round up their sheep. The local wildlife bureau should be able to direct you to the best areas for viewing.

Argali in the mountains of Gansu province – a good place to spot them.

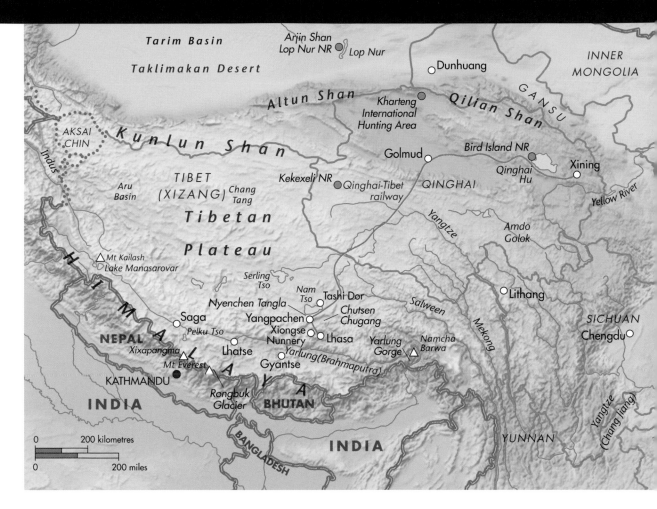

birds, Qinghai Lake (Koko Nor)

China's largest lake, at 3200 metres (10,500 feet), is venerated by Tibetans as the celestial lake Koko Nor – 'azure blue sea'. From May to September, the lake is a focal point for 190 bird species, and in May, thousands of bar-headed geese, cormorants and two species of gulls nest on one of several 'islands' within Niao Dao Nature Reserve – a peninsula on the northwestern shore. There is an underground viewing area on Egg Island, but expect crowds at peak times – worst during the week of the May

holiday. Take the Qinghai-Tibet railway from either Beijing or Lhasa, getting off at the new Qinghai Lake station, or fly to Xining and take a four-hour bus or taxi ride towards Golmud. En route, it is common to see pilgrims prostrating themselves around the lake's shoreline on a holy Koran. The round trip can be booked from many hotels in Xining.

kiangs (wild asses)

Large herds do still occur though generally only in remote areas, but an area to look for them not too far off the beaten track is north of

Xixapangma massif. When heading from Lhatse towards Nepal, turn off the Friendship Highway at the bottom of Lalung La towards Saga and Pelku Tso – an azure lake worth a visit. The drive takes several hours (road conditions allowing), and you may see kiangs and Tibetan gazelles on the way. Watch for migrating eagles hunting pikas.

pikas

Late September is a good month for pika-watching, when these small relatives of rabbits are

A wild yak in defensive-threat pose before fleeing. Poaching has made wild yaks very nervous.

building up winter stores and are particularly active – snow cover also reveals their burrows. A good location is north of Xixapangma base-camp, just off the Friendship Highway – a two-day drive from Lhasa. The area beside the dirt track which heads towards Saga is dotted with pika burrows, and an hour's watch should reward you with foraging behaviour and possibly an attempted eagle kill.

wild yak

The best place to see these elusive, high-altitude beasts is Yeniugou ('wild yak') valley in western Qinghai, a few hours south of Golmud. You will need a guide (not easy to find) and a 4x4 to navigate through the valleys. Field skills are necessary even to have a chance of seeing a yak, and you'll need to be fit to hike at this height.

Keep your eyes peeled for brown bears. At this altitude, they live almost solely on pikas, as vegetation is almost non-existent for nine months of the year. Avoid the coldest months (November to March) and the summer rains (July and August). May and June are best, when temperatures are agreeable but rain hasn't made the ground impassable. Be aware of altitude – as the ascent from Golmud is pretty sudden and needs to be factored into your itinerary.

LANDSCAPE WONDERS

hot springs

The plateau is dotted with hot springs – a reminder that the Himalayas are still growing. The most famous is at Yangpachen, a settlement 120km (76 miles) from Lhasa. Here you can bathe while contemplating how water boils at 4850 metres (15,910 feet) above sea level. Accommodation is basic and requires prebooking, and so bring a tent as backup.

In summer, combine a cultural experience and bathing with snakes at Chutsen Chugang hot springs by Tidrum Nunnery, Mozhugongka county – a three-hour drive east of Lhasa, with basic accommodation nearby.

Nam Tso

This is one of the highest saltwater lakes in the world, at just above 4700 metres (15,420 feet), and is second only in size to Qinghai Lake. When free of ice in summer, its azure-blue waters are a Mecca for migratory birds. Melt-water from the nearby Nyenchen Tangla range supports lush grass, frequented by nomadic pastoralists and pilgrims.

The azure, saltwater Nam Tso.

You can ride yaks around the lake, visit the sacred stupas (Buddhist memorials) at the Tashi Dor monastery or just take in the breathtaking scenery.

Nam Tso is 180km (112 miles) north of Lhasa. Follow the Qinghai-Tibet railway north, turn off the main road towards Damxung and cross the Nyenchen Tangla (a two-hour drive) over the Largen La (a pass blocked with snow in winter). Keep an eye out for bearded vultures scouring the mountains and lake edges for carrion. Accommodation is so basic that camping is the best option.

Everest advance base-camp

Everest, or Qomolangma – the Mecca of world mountaineering – is one of nine 8000-metre (26,240-foot) peaks in Tibet. The two weather windows offering relatively safe temperatures and clear skies are in May and September, though snowstorms and high winds can hit at any time. There are tours from Lhasa (and Nepal), but go only if you are a fit and regular hiker – and be vigilant for symptoms of altitude sickness. You can use yaks to carry your equipment while trekking along the Rongbuk Glacier. The reward is to reach altitudes of 6400 metres (4 miles) above sea level, with stunning views of Everest and other Himalayan peaks.

Thong La

Travel south on the Friendship Highway, past Tingri (and Mt Everest) towards Nepal up to 5150 metres (16,900 feet) at Thong La. This pass offers possibly the best views of the grand barrier in China. The road to Nepal then turns into the longest downhill ride in the world – unmissable.

driving off the plateau

Thought to have inspired the notion of Shangri-La, the Yarlung Gorge is the deepest valley in the world, with climatic zones ranging from arctic to subtropical.

Described as the 'Everest of rivers', the Yarlung, which originates from Mt Kailash and eventually becomes the Brahmaputra, drops 3000 metres (9840 feet) in 242km (150 miles).

The impenetrable gorge allegedly hides Tibetan pygmy tribes, some of the largest trees in Asia and many species of animals and birds that are almost extinct elsewhere.

Trips to the gorge are almost impossible to organize, but you can still drive off the edge of the roof of the world by following the Friendship Highway to Nepal's capital Kathmandu.

As you start the descent of the longest downhill on Earth, you are flanked to your right by the snow-capped Xixapangma massif, one of nine 8000-metre (26,240-foot) mountains in Tibet. In three hours, you'll drop four vertical kilometres, cutting through the largest mountain range on Earth. After an

The longest downhill on Earth, winding down the Friendship Highway to Nepal.

hour, the gorge narrows and snow gives way to endless waterfalls that cascade over a hillside cloaked with conifers. As the gorge plunges ever downwards, the air becomes heavier and wetter, and you'll become more and more lethargic as temperatures (and oxygen levels) soar, the birds start to sing, and clouds appear out of nowhere as they battle to climb the gorge. A brief stop at the border allows you to enter Nepal, where a rainbow of colour awaits. It's a fantastic experience to end what will have been the trip of your lifetime – and will make you realize that we are just not designed to live at altitude.

CULTURAL WONDERS

Potala Palace, steeped in Tibetan history, religion and art, and former home of the Dalai Lama.

bu collection
In mid-May, the high-altitude pastures of eastern Tibet bloom, attracting thousands of nomads to search for 'summer grass, winter worm', or bu – the performance-enhancing fungus that grows from the heads of ghost moth caterpillars. Tented villages spring up in the middle of nowhere – pool tables sit out on grassy meadows, smoke rises from yak-hair tents. Occasionally the concentrated search is punctuted by a high-pitched wail as a caterpillar is found.

Amdo Golok in Qinghai province is famous for bu collection. Here you can drink yak-butter tea with nomads in their tents, search for bu yourself and visit local (relatively!) towns where it is sold. Fly to Xining and then drive to Golok – a two-day journey but worth it.

Tours are best arranged in advance, and you'll need a good guide. Another possible collecting area to visit is Nagchu in Tibet.

Potala Palace
No trip to the capital city of Tibet will be complete without visiting the Potala Palace, visible from almost everywhere in Lhasa. The former home of the exiled Dalai Lama is now more of a museum than a seat of government and place of worship. Perched on Marpo Ri Hill, the palace is possibly the most impressive building in Tibet. (Take the steps slowly if this is your first experience of exertion at altitude.) Early legends tell of a sacred cave that was used as a retreat by Emperor Songtsen Gampo in the seventh century AD and probably led to the creation of the palace. Walk through the halls and wonder what life was like when it was lived in, while marvelling at the treasure trove of Tibetan history, religion, culture and art.

Jokhang Temple, dawn
Built in AD 647, Jokhang is one of the oldest surviving buildings in Lhasa and probably Tibet's most holy temple – a focus for Buddhists the world over. At dawn, juniper twigs fuel the giant incense burners that stand in front.

Perfumed smoke billows from them, and occasional bursts of flames illuminate a ghostly scene of worshippers. The murmuring of sacred mantras and the rhythmical swish of hands against cold stone, mixed with the aroma of juniper is a combination that will take a lifetime to forget. There is a perpetual flow of pilgrims of all ages shuffling clockwise around the temple and prostrating themselves, repeating prayers as they touch their foreheads on the cold stone.

October is a good time to visit Lhasa – temperatures are relatively warm and skies clear, and there is an influx of nomadic pilgrims. It is also a good time to embark on organized trips outside the city, but prebook, as it's a busy period.

Xiongse Nunnery

The relationships between people and animals are fundamental to the Buddhist religion, and nowhere is this seen better than at monasteries and nunneries where devout Buddhists feed wild animals through the cold Tibetan winters. Xiongse Nunnery in Caina village, founded in 1181, is just 30km (19 miles) from Lhasa (towards Gyantse), but it will take two hours to walk up to the temple. In winter, nuns feed 25 or so Tibetan-eared pheasants in the early morning.

Rongbuk Monastery

The highest monastery in the world would be worth a visit in any location, but with Everest as its backdrop, it's doubly special. Rongbuk is a 20-minute jeep ride or a two-hour walk from Everest base-camp on the Friendship Highway and has an adjacent guesthouse. The weather is best in May or September and October.

Saga Dawa Festival, Mt Kailash

The 1000km (620-mile) journey from Lhasa to what many regard as the most holy place on Earth requires a five-day drive on dirt roads. Ensure your driver goes at a sensible speed – crashes are common – and bring snacks and an inflatable sleeping mat, as you will be camping for most of the way (a cleaner alternative to many of the basic guesthouses). The route follows the grand barrier of the Himalayas, giving tantalizing glimpses of snow-capped peaks, and you'll see pilgrims prostrating themselves along the road for many miles before Mt Kailash is even visible.

Visit Mt Kailash at full moon in the fourth lunar month of the Tibetan calendar (usually June) to witness the spectacle of the Saga Dawa Festival. Throngs of worshippers camp around Kailash and undergo the 55km (33-mile) kora, walking or prostrating themselves around the mountain.

Walking the route takes about three days, but you can hire yaks and porters. Almighty roars accompany the climax of the festival – the raising of the prayer-flag-festooned flagpole. Not to be missed.

Tibetan-eared pheasant, one of many fed year-round by the nuns at Xiongse Nunnery.

4 yunnan

If you don't have time to visit more than one region, visit Yunnan. It is unrivalled in its range of landscapes – from snow-covered mountains to steamy rainforests – and its diversity of people and wildlife. It has a third (25) of China's ethnic minorities, half of the country's plant and animal species and more than 100 nature reserves – far more than any other province. The climate varies hugely, but the capital Kunming has a pleasant climate year round. Tourism is booming, which has its downside but also means travel is relatively easy.

WILD WONDERS

Wild Elephant Valley – sightings are regular.

Asian elephants

Wild Elephant Valley in Sanchahe Nature Reserve, Xishuangbanna, is a tourist hotspot but offers the chance of a fairly close encounter. A 2km (1.2-mile) cable-car ride takes you to the forested viewing area – high walkways and viewing platforms above the river where the elephants come to bathe early in the morning. To get the best chance of seeing the elephants, go in spring and stay overnight in one of the basic canopy treehouses. Or you can get your photo taken by the river and have the elephants Photoshopped in behind you – a service offered on-site for a small price! Unfortunately, many of the tourists visiting Wild Elephant Valley will be shuttled in to watch the depressing 'wild' elephant performances – not to be recommended if you care about animal welfare.

black-crested gibbons

Once widely distributed in China, from the east coast to the border with Burma, black-crested gibbons are now restricted to remnant patches of forest in Yunnan, southwest Guangxi and one small population on Hainan Island. There are thought to be 200–400 gibbons in Wuliang Shan Nature Reserve in the Wuliang Mountains, but seeing them is not easy, as they live in family groups and spend virtually all their time foraging high in the canopy. Male black-crested gibbons are black and females are yellow. At first light most mornings, they will duet from the treetops, which is a good time to pinpoint their location. Staking out fruiting trees in spring is another way to get a sighting. The closest airport is Kunming-Dali, a full day's drive from Wuliang Shan.

pheasants

China has 27 species of pheasant – more than half the world total – some of the most beautiful of which can be found in the southwest. They are incredibly shy and difficult to see, though birding tours can take you to the top sites: Lady Amherst's in the Cang Shan (Jade Green Mountains), for example. Visit in May when the climate is pleasant. Neighbouring Sichuan province probably has some of the most accessible birding reserves, including Wolong, 134km (83 miles) from Chengdu, which has Temminck's tragopan, white-eared pheasant, golden pheasant and Chinese monal. To see large numbers of white-eared pheasants, travel from Chengdu to Kanding and a day's journey on to Daocheng, to Zhujie monastery, where they are hand-fed by Tibetan monks.

Lady Amherst's pheasant, a Yunnan gem.

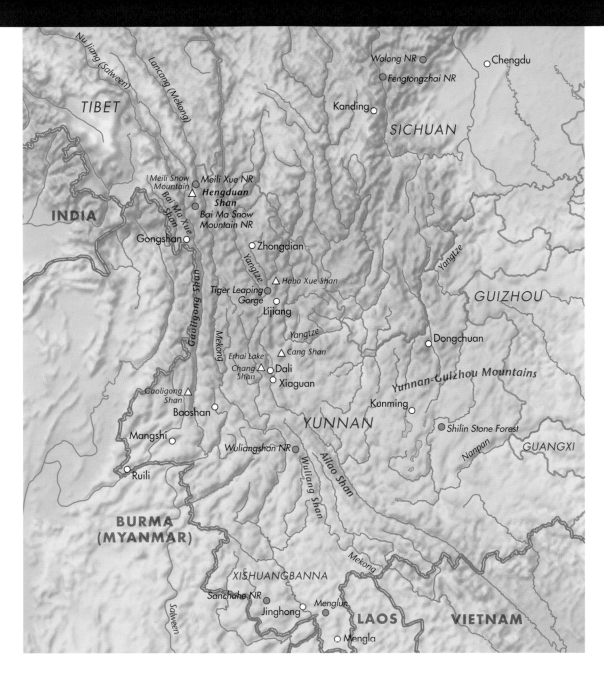

The map shows the region including Tibet, India, Sichuan, Guizhou, Guangxi, Yunnan, Burma (Myanmar), Laos and Vietnam, with labelled locations:

Nu Jiang (Salween), Lancang (Mekong), Wolong NR, Chengdu, Fengtongzhai NR, Kanding, SICHUAN, TIBET, Meili Snow Mountain, Meili Xue NR, Hengduan Shan, Bai Ma Xue Shan, Bai Ma Snow Mountain NR, INDIA, Gongshan, Zhongdian, Yangtze, Haba Xue Shan, GUIZHOU, Gaoligong Shan, Tiger Leaping Gorge, Lijiang, Mekong, Yangtze, Dongchuan, Erhai Lake, Cang Shan, Chang Shan, Dali, Yunnan-Guizhou Mountains, Xiaguan, Gaoligong Shan, Baoshan, Kunming, YUNNAN, Mangshi, Wuliangshan NR, Shilin Stone Forest, GUANGXI, Nanpan, Ruili, Wuliang Shan, Ailao Shan, BURMA (MYANMAR), Mekong, Salween, XISHUANGBANNA, Sanchahe NR, Menglun, Jinghong, LAOS, VIETNAM, Mengla

plant paradise: Zhongdian

Yunnan has the richest diversity of plants in China, with more than 15,000 species. For botanical delights, take a tour to Zhongdian in the north, otherwise known as Shangri-La. Fly to the Tibetan settlement of Zhongdian. It sits on a 3000-metre (9840-foot) plateau dotted with villages of ornamental wooden houses and surrounded by mountains and forests. It was in this region that the plant hunters Joseph Rock, George Forrest and Frank Kingdon-Ward collected seeds and cuttings from trees, shrubs such as rhododendrons and azaleas and herbaceous perennials

One of many primulas originating in Yunnan.

Xishuangbanna Tropical Botanical Garden – Menglun

Founded in 1959 by the famous Chinese botanist Cai Xitao, who was fascinated by the medicinal qualities of the region's tropical forest plants, this wonderful garden in Menglun town, east of Jinghong, has 860ha (2125 acres) displaying thousands of different tropical and subtropical plants. Highlights include the 800-year-old tea tree, 1000-year-old cycads, bamboo groves and orchid gardens.

Yunnan snub-nosed monkeys

These little-studied monkeys live at higher altitudes than any other primate except humans, and seeing them is truly magical. Only 1500 remain in the wild. Tacheng in Bai Ma Xue Shan (Bai Ma Snow Mountain) Nature Reserve has the largest population. If possible, go with an experienced tour guide. Fly to Zhongdian and travel by mountain road for two hours to Tacheng. One group is frequently herded by the reserve staff into a smaller area of forest –

controversial but gives tourists a better chance of seeing them.

LANDSCAPE WONDERS

Meili Xue Shan

Straddling the Yunnan-Tibet border, the Meili Snow Mountains divide the deep gorges of the Mekong and Nu Jiang (Salween) rivers and include Kawagebo – the tallest mountain in Yunnan, at 6672 metres (22,241 feet), and one of the eight sacred mountains of Tibetan Buddhism. A view of Kawagebo from Feilai Temple – 10km (6 miles) from Deqin – is worth a sunrise trip. From there it's 20km (12 miles) to the Meili Xue Shan Nature Reserve. Hike to the Mingyong glacier in the reserve or trek to the waterfalls and hot springs via Tibetan settlements. There is a 12-day pilgrimage trek around Kawa Karpa (you'll need a guide). Fly to Zhongdian and catch a bus or hire a car to Deqin (heavy snow from December to April may close the roads).

that are now classic garden plants. In June, the meadows are full of flowers including primulas and irises. Lakes Napha and Sudu, both within two hours' drive, offer plants such as blue poppies and slipper orchids, as well as woodlands carpeted with flowers.

red pandas

A few reserves in southwest China claim to have red pandas, with Fengtongzhai Nature Reserve – about 250km (155 miles) from Chengdu, the capital of Sichuan – offering the best chance of seeing them. Go in spring. But since red pandas are normally solitary and active from dusk to dawn, you'll need good binoculars, patience and a lot of luck. Chengdu Panda Breeding and Research Centre, 10km (6miles) from downtown Chengdu, has captive red pandas and so does Kunming's safari park.

Red pandas – difficult to see in the wild but on view at the Chengdu breeding centre.

Tiger Leaping Gorge, or Hutiao Xia, one of China's tourist hotspots.

Gaoligong Shan and Nu Jiang Gorge

The Gaoligong Mountains mark the border of Yunnan and Burma, forming a divide between the Nu Jiang (Salween) and the Irawaddy rivers. The Nu Jiang has carved a 805km-long (500-mile) gorge alongside the mountains – a beautiful, rugged area. These mountains, their magical forests and the gorge are a global biodiversity hotspot and epicentre of plant endemism (uniqueness) in northwest Yunnan. This is probably Yunnan's last true wilderness. The best way to explore the wild Nu Jiang Gorge is to join a guided trek from Dali, Baoshan or Lijiang. Clinging to the gorge sides are Tibetan, Lisu and Nu settlements, their precarious rope bridges dangling over the rapids. Construction of 13 dams on the

Nu Jiang in Yunnan was halted in 2004, but the Thai and Burmese governments still plan to build several, including the Ta Sang Dam, said to be larger than the controversial Three Gorges Dam.

Tiger Leaping Gorge

This incredibly deep canyon, advertised as one of China's great natural wonders, is fairly commercial now. It is about 90km (60 miles) north of Lijiang in northwestern Yunnan, and runs between Jade Dragon Snow Mountain (Yulong Xue Shan) and Haba Snow Mountain (Haba Xue Shan). From the cliff-edge, you can look down 200 metres (655 feet) onto the raging Yangtze, known locally as the Jingsha, or Golden Sands River. Visit in May and June for the spring flowers.

Allow three days to hike along the gorge. Or catch a bus to Qiaotou, from where buses will take you to viewpoints. It is very touristy and doesn't necessarily match up to the hype. Eight dams are being planned on the upper reaches of the Yangtze, threatening this gorge – part of a World Heritage Area – and displacing thousands of Naxi people. See it before it goes.

Erhai Lake

The huge 'ear-shaped lake', just north of Dali city, is one of the headwaters of the Mekong River. From Dali, walk there in an hour or

Fishermen on Erhai Lake, still using trained cormorants, and not just for tourist displays.

cycle there in ten minutes and then pedal leisurely around the lake. Fishermen here still fish with cormorants. The surrounding villages and islands are worth visiting. Ferries will take you across the lake for early-morning markets or to visit one of the pavilions on the east side. To stay overnight, take the afternoon ferry from Caicun to Wase.

Mengla rainforest, Xishuangbanna

The tip of Yunnan, bordering Laos and Burma, boasts China's only tropical rainforests, now in the Xishuangbanna Biosphere Reserve. Occupying less than 0.2 per cent of the country, this area has the richest biodiversity in China: 4000 plant species, 102 mammal species and 400 bird species, not counting the numerous reptiles, amphibians, fish and invertebrates. Much of the

forest has been cleared for rubber plantations, but pockets remain. The best are the sub-reserves Menglun and Mengla (the largest).

Fly to Jinghong from Kunming and drive to the town of Mengla. About 7m (4 miles) north of Mengla at Bupan is a canopy walkway above the forest.

Shilin Stone Forest

About 126km (78 miles) southeast of Kunming is a bizarre collection of limestone pinnacles – Shilin Stone Forest. Over millions of years, the ancient seabed has been attacked by wind, water and acid rain, resulting in the 'forest' of grey stone pillars. It's one of Yunnan's top tourist attractions, but if you have visited some of Yunnan's wilder wonders, you may feel that it's a little overrated. The Sani people (a branch of the Yi tribe) live here.

CULTURAL WONDERS

Dai water-splashing festival

As part of their new-year celebrations, the Dai people of Xishuangbanna hold a water-splashing festival in April (dates change). It symbolizes the washing away of the dirt and bad luck of the past year and welcomes in the happiness of the new and a good wet season. See dragon-boat races on the Lancang River on the first day (there is a huge market, too). Experience more intimate celebrations with the washing of Buddha effigies in the local village temples. In Jinghong, be prepared to get wet in the following days as the biggest water fight of all time (it's good luck to get really wet) takes place in the streets. Look out for the Dai peacock and elephant drum dances. The finale is best viewed beside the river – the site of a huge firework display.

minority villages Xishuangbanna

This is the most ethnically diverse area in China. Among the 880,000 population are Dai, Ahka, Jinuo, Yi, Yao, Lahu, Bulanand and Yao, to name but a few, who have retained religions, cultures and languages in common with adjacent Burma, Laos, Thailand and Vietnam. Base yourself in Jinghong, and spend a few days trekking to villages. With the development of tourism, traditional practices and clothing are disappearing, and in some

The limestone pinnacles of Shilin's Stone Forest – a major tourist area.

Dai men and women rowing a dragon boat at the water-splashing festival in Xishuangbanna, which takes place in April or May.

villages, they are on display just for tourists.

Lijiang old town and the Naxi

About 160km (100 miles) north of Dali, in a valley below the Yulong Xue Shan (Jade Dragon Snow Mountain), lies Lijiang. Hire a bicycle and soak up the Naxi culture. If you get past the tacky attractions that have gradually made much of Lijiang a tawdry tourist spot, you will find genuine culture in the cobbled streets, the winding canals, the rickety wooden buildings and the hubbub of the old town's bustling markets. The Naxi are descended from Tibetan Quiang tribes, who settled in the region 1400 years ago. Their religion is Dongba, whose scriptures and murals can be seen on the walls of the temples in town and the surrounding villages. Head to Yufeng Si monastery for a

magnificent view of the northern Lijiang valley. Attractive villages and monasteries such as Puji and Fuguo set among beautiful scenery can be explored by bicycle. Further afield is Yuhu, the village where Joseph Rock – famous botanist, explorer and champion of the Naxi people – lived in the 1920s and 1930s.

Dali and the Bai

The laid-back town of Dali lies on the western edge of Erhai Lake under the gaze of the Cang Shan (Jade Green Mountains). It is home to about 1.5 million people from the Bai minority, who have lived in the region for 3000 or more years. Fly from Kunming to Xiaguan, from where it's a 45-minute drive. The Third Moon Fair, originally a Buddhist festival, on the fifteenth of the third lunar month (April or May) – is five days of festivities and markets that attracts people from

all over Yunnan. Dali still has pagodas from the time when it was a Buddhist centre, the best being San Ta Si (the Three Pagodas), just north of town. Excursions to Cang Shan can be booked from Dali.

Ruili and minority villages

The town is on the border with Burma and is an entrepôt for many goods. It's also culturally fascinating, with a mix of Han minorities and Burmese, who are the main market stallholders. Visit the market in the west of town in the morning and the one in the east later, especially if you want to buy jade.

Traditional villages and temples are accessible by bike and are not as oversubscribed as those in Xishuangbanna. You can also fly to Ruili from Mangshi, which is two hours from Kunming.

5 the great rice bowl (south China)

Southern China offers everything from sacred mountains, uncharted caves and rare wildlife to traditional rural landscapes and ethnic cultures. Spring is the best time to visit, as summer can get very humid and autumn can be very wet. Travel is fairly cheap, and most areas can be reached by air and road. The environment suffers from intensive cultivation, industrial waste and pollution, and be prepared for litter and bad air-quality. But outside the cities, you will find areas of outstanding natural beauty and plenty of national nature reserves.

WILD WONDERS

birdwatching – Poyang and Caohai lakes

Poyang in Jiangxi province in the central Yangtze floodplain is China's biggest lake, an international Ramsar wetland site and arguably the largest migratory habitat on the planet, leaving even the hardiest twitcher dumbstruck at the millions of birds that flock here from December to March. It is famous as the wintering ground for Siberian cranes.

Other rare and endangered species include the oriental white stork, swan goose and white-naped crane. Seeing them close isn't easy, and so allow plenty of time and take a powerful telescope.

Caohai Lake is much smaller but also a premier birding site. The nature reserve is a 15-minute drive from central Weining in western Guizhou. Here you can see overwintering endangered black-necked cranes, along with more than 180 other species.

black-headed leaf monkeys

Black-headed leaf monkeys, also called Francois' langurs, are among southern China's most endearing and endangered monkeys. The remote Mayanghe National Nature Reserve in northwest Guizhou has

White-naped cranes wintering at Caohai.

possibly the largest population – about 700, including habituated groups. The closest town with an airport is Tongren, where you can pick up a bus to take you to Yenhe and on from there to Mayanghe – a long, queasy day's drive on winding mountain roads. Accommodation is basic. Rise early and accompany the guides for a ten-minute stroll beside the river to the feeding station. A few long blows on a whistle and the temptation of sweet potato will bring a group of monkeys down the steep walls of the gorge to be fed. To see more, take a taxi up the valley, where other groups in the reserve can be

seen fairly easily. Visit between February and April to see the orange babies.

Chinese alligators

With fewer than 150 left, you are unlikely to see Chinese alligators in the wild. Instead, visit the Anhui Alligator Breeding Centre at Xuancheng, about an hour's drive southeast of Wuhu; travel companies at Tunxi or Heifei can arrange visits. The 500 or so adults are kept in natural-looking breeding ponds, and hundreds of smaller alligators are kept in brick-lined ponds. Or see them at Changxing Breeding and Research Centre in the

Farm-bred baby Chinese alligators.

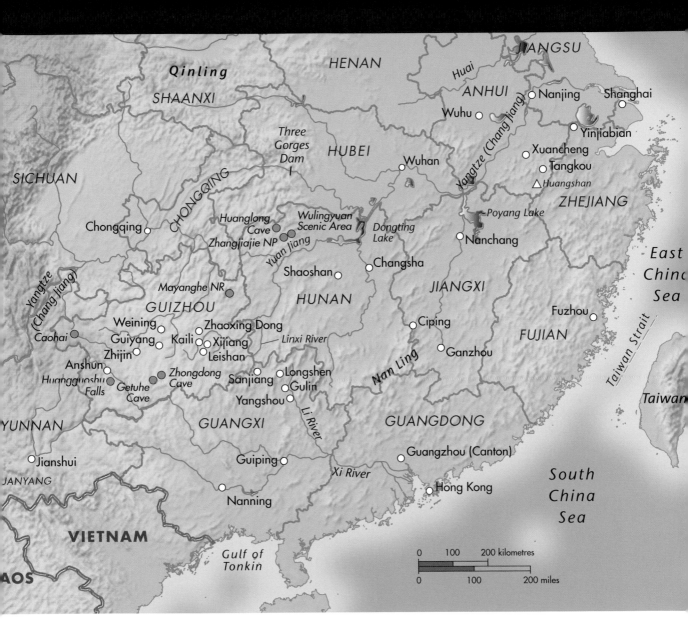

village of Yinjiabian, in Changxing county, Zhejiang province. In May or June, the alligators are courting, and you may hear the bellows of males calling in the early morning.

cormorant fishing

Fishing using cormorants is still fairly common in central and southern China, though often just for the benefit of tourists. The Li River in northeast Guanxi province offers the most picturesque setting. Tours run from hotels and cafés in Yangshuo, near Guilin, punting you out on the river at dusk to watch the performance of cormorants fishing from a bamboo raft.

Huang Shan's macaques

If you visit the famous Huang Shan (Yellow Mountain) scenic area, go to Monkey Valley – about an hour's drive from Huangshan airport, between Tunxi and Tangkou. Immediately after a tollgate and at the entrance to Zhaixi village, a small road leads west towards the

wild china 237

village of Fuxi, where you can drive to the entrance gate of Monkey Valley. It is about a 15-minute walk up steps to the viewing location. Here, several groups of about 40 of these very large macaques are fed 4 times a day. Watch from the safety of the wooden pagodas, but obey the rules of monkey etiquette – don't stare into a monkey's eyes, or you could provoke an angry charge.

wild-animal markets

If you want to grasp the full scale of trade and consumption of wildlife in China, take a trip to one of the many live-animal markets of the southern provinces for an eye-opening but distressing experience. The meat and body parts of some animals are believed to have medicinal or aphrodisiac qualities, but many others are simply eaten in restaurants to prove the host's wealth and status. One of the most notorious markets is Hua Nam on the Zhengcha Road in Guangzhou.

LANDSCAPE WONDERS

karst landscape and the Li River

From the famous town of Guilin in Guangxi province, the Li River meanders southward through quintessential Chinese scenery, as depicted in so many watercolours, textiles and artefacts. Contorted karst peaks, each with its own name and legend, flank the river, and between them is a patchwork of green fields. The scenery is at its best 65km (40 miles) south at Yangshuo. Hire a bike and pedal among rice fields surrounded by the green-topped limestone peaks, or take a boat ride along the Li from Yangshuo or Guilin to the

The amazing karst of Guangxi's cave country.

densest concentration of peaks between the villages of Caoping and Xingping. The landscape is at its most luxuriant between May and September.

Yuanyang rice terraces

Some of the most magnificent rice terraces in China, stretching as far as you can see, are the work of the Hani in Yunnan province. Base yourself at the hill town of Yuanyang, 80km (50 miles) south of Jianshui. The fields are clouded in mist in spring and summer and are at their best visually in winter, when they are flooded. It's well worth a sunrise or sunset vigil for incredible photo opportunities. Locals can direct you to the best spots. The famous Dragon's Back rice terraces in Longshen in Guangxi are as impressive.

Live snakes for sale. Markets give an indication of the scale of wild-animal consumption.

Zhangjiajie

The landscape of the Wulingyuan Scenic Area, often referred to as Zhangjiajie, is among the most mystical and dramatic you are likely to see. This UNESCO World Heritage Area, consisting of Zhangjiajie National Park, Suoxiyu Nature Reserve and Tianzi Mountain Natural Reserve, is in northwest Hunan.

Its sandstone shelves are covered with more than 3000 limestone towers soaring 300 metres (985 feet) above a misty, subtropical forest scored by tumbling waterfalls and crystal streams. The region boasts more than 3000 plant species, including some 550 trees, and rare animals such as giant salamanders, though very few remain.

Several towns serve as access points, but Zhangjiajie village at the southern boundary tends to be the most popular. Or stay at Suoxi village and explore the east and north of the reserve. It's humid in summer, and it can snow in winter.

Yellow Dragon Cave and Zhijin caves

Unless you are a very experienced caver and can hire a knowledgeable local guide, then exploring China's caves is a dangerous undertaking. The best starting point is to contact one of the British caving groups that have explored some of them or the karst institute in Guizhou. For a less risky trip into the underworld, visit one of China's remarkable show caves, such as Yellow Dragon Cave (Huanglong Dong), close to Zhangjiajie in Hunan province, or Zhijin in Guizhou province – ranked as China's largest (though many are still unexplored). Zhijin can be reached by bus as a long day-trip from Anshun. The caves are gaudily lit and a bit tacky but have stairways, walkways and boats to take you safely on a tour of the geological phenomena.

Huang Shan (Yellow Mountains)

It is claimed that, once you have visited the Huang Shan, you will never want to climb another mountain. Its spectacular panorama – monoliths of rock rising up from bamboo forests, and ancient, contorted pines clinging to rock faces that sit among swirling mists – has attracted poets and artists for hundreds of years. The many steps to the various viewing points on Huang Shan can be busy. If you can't manage the climb, porters can take you in a sedan

The sandstone pillars of the inspirational Wulingyuan Scenic Area.

The famous rocks of Huang Shan rising above the clouds – a tourist site worth the crowds.

chair. To see the sun rise over the mountains, stay in a mountain hotel. But you'll still need to get to a viewing point early to get an ideal spot. The main gateway to Huang Shan is Tangkou, which can be reached by bus from Tunxi – accessible by rail and air. To experience the sea of clouds, it is best to go in November. Or visit in midwinter when the mountains are cloaked in snow. Take warm clothes and a waterproof, and be prepared to queue for a cable car.

Huangguoshu Falls

This is the largest waterfall in Asia – more than 80 metres (265 feet) wide, with a thundering 74-metre (245-foot) cascade down into Rhinoceros Pool. It can be visited in a day trip from Anshun in western Guizhou. Take a cable-car ride to the entrance just before Huangguoshu village, and explore the falls on a pathway over the pool or through a tunnel behind the water curtain. Take waterproofs and strong footwear.

CULTURAL WONDERS

Miao and Dong villages

The Qiandongnan Miao and Dong autonomous prefecture in southeast Guizhou is home to more than 13 different minorities maintaining a traditional way of life. Kaili (west of Sanjiang) is the centre of the Miao silver culture and a good place to base yourself. Visit Chong'an, a small village by the river, two hours north of Kaili, and the picturesque Xijiang, which is the largest Miao village in the region and holds a weekly market famous for its embroidery and silverware. Head southeast of Kaili

A Miao woman in traditional silver headwear.

to the Dong village of Zhaoxing, with its traditional wooden houses, wind-and-rain bridges and five-drum towers. The towers were lookout posts – at a sign of danger, drums in them were beaten to rouse the villages. Visit in spring for the famous Lusheng Festival.

Sanjiang wooden bridges
The most famous of the wind-and-rain bridges is in Guangxi province, north of Sanjiang in Chengyang, on the far side of the Li River. Not a single nail was used in the construction of these bridges, which once served a religious purpose but are now where the locals gather to gossip under their eaves – great places to sit and take in the atmosphere.

Shaoshan – birthplace of Mao
Chairman Mao Zedong was born in 1893 in the tiny town of Shaoshan. It is a three-hour journey by train from Changsha, in a pretty area of mountainous countryside, scattered with rice paddies and traditional houses – but a busy tourist spot. The shrine is Mao's family home, a simple mud-brick house, but there is also the Museum of Comrade Mao and the Mao ancestral temple.

'Cradle' of the revolution
Mao led 900 men to Mt Jinggang in 1927, and it was from here, on the Hunan-Jiangxi border, that he launched the famous 'long march' into Shaanxi in 1934, with Zhou Enlai and Zhu De, that established the People's Republic of China.

The hills are peppered with more than a hundred Red Army historical sites. In Ciping, which you can reach by bus from Nanchang and Ganzhou, they include the revolutionary museum and the former revolutionary headquarters.

You can also escape into the surrounding countryside, where tourists are scarce and the highland forests boast square-stemmed bamboo and alpine azaleas. The best time to go is between June and October.

A wooden wind-and-rain bridge at Guangxi, made by the Dong without the use of a single nail.

6 crowded shores

The bustling modern cities of Shanghai and Hong Kong are not the only places to visit along China's coastline. Though this is one of the most densely populated and highly exploited coastlines on Earth (700 million people live here), it stretches 14,500km (9010 miles) from the frozen north to the tropical south and offers a diversity of landscapes. The pockets of reedbeds, mudflats and marshes that remain are a must for birdwatchers, especially in winter, when millions of migrants, including some of the world's rarest birds, refuel or overwinter there.

The endangered black-faced spoonbill. Deep Bay is one of only three wintering sites for it.

WILD WONDERS

birds – Mai Po Marshes

The Mai Po Marshes Nature Reserve, run by WWF, is Hong Kong's top birdwatching site, and the marshes (originally fish ponds) and shallow estuary of Inner Deep Bay in the northwestern corner of Hong Kong are a Ramsar Wetland of International Importance. From November to March, overwintering migrants include the oriental stork, Saunders's gull and black-faced spoonbill. In spring, there are fewer birds but lots of dragonflies and butterflies. There are hides, boardwalks, guided tours, nature trails, a field centre and a museum. Travel there from Sheung Shui or Yuen Long by bus or taxi (you can stay in dormitories at the Peter Scott Field Studies Centre). There are also public tours.

botanic garden – Hong Kong

Kadoorie Farm Botanic Garden is an important conservation and educational centre lying below Kwun Yum Shan (Goddess of Mercy Mountain) in Hong Kong's central New Territories. It's also a place to escape the oppressive city. More than1000 native plant species are on display. The garden also has numerous nature trails and exhibits, including an amphibian and reptile house, a butterfly garden, wildfowl ponds, a stream-life display and a wild-animal rescue centre. Take the KCR train to Tai Po Market or Tai Wo and the 64K bus towards Yuen Long (the last stop on the line).

chinese white dolphin

Little is known about the Chinese white dolphin, but it is undoubtedly under threat from habitat loss, pollution, overfishing and an increase in marine traffic in the waters of Hong Kong. The best way to see the dolphins is on a Hong Kong Dolphinwatch tour in the waters about Lantau. It is also possible to charter a boat, but keep in mind that there are strict limits to how close you can approach.

macaques – Hainan Island

About a thousand rhesus macaques inhabit the Macaque Nature Reserve on the Nanwan peninsula in Lingshui county, near Xincun. Take a ten-minute ferry crossing from Xincun pier and a taxi to the visitors' centre. If you walk, you'll get good views of wild monkeys. There is also a cable car that runs between Xincun and monkey island. The monkeys at the visitor centre are used to people, but leave any food you have in the lockers so as not to tempt them. There are depressing shows featuring trained monkeys. Opt for

Map labels:

MONGOLIA

INNER MONGOLIA

GOBI

Yellow River (Huang He)

Hohhot

BEIJING

HEBEI

SHANXI

Shijiazhuang

SHAANXI

Wei

Yellow River

HENAN

Three Gorges Dam

HUBEI

Wuhan

Nanchang

Changsha

HUNAN

GUIZHOU

GUANGXI

Nanning

Guangdong

Yuen Long

Mai Po Marshes Nature Reserve

Gulf of Tonkin

Haikou

Hainan

Five Fingers Mt

nfengling NR

Sanya National Coral Reef NR

Bawangling NR

Nanwan

Yalong Bay

Qiqihaer

Zhalong NR

Harbin

HEILONGJIANG

Manchurian Plain

Liao River

JILIN

Shenyang

Shuangtai He

Shuangtai Hekou NR

Liaodong Bay

Yinkou

NORTH KOREA

Shedao Island

Dalian

Bo Hai Gulf

Yangtai

Weihai Airport

Penglai

Chengshan

Shandong Peninsula

Rongcheng

SOUTH KOREA

Sea of Japan

Jinan

Tai Shan

SHANDONG

Yellow Sea

Mt Jinping

Yancheng NR

Dafeng NR

JIANGSU

Huai

Nanjing

ANHUI

Qing Pu

Shanghai

Chongming Island

Huangpu River

Hangzhou

Yanguan

Hangzhou Bay

Qiantang River

West Lake

ZHEJIANG

East China Sea

Yangtze (Chang Jiang)

Poyang Hu

Wuyishan NR

Wuyigong

JIANGXI

Fuzhou

Putian

FUJIAN

Yongding

Xiamen

Taiwan Strait

Taiwan

GUANGDONG

Guangzhou

Shenzhen

Hong Kong

Lantau Island

Pearl River

Dongsha Islands

South China Sea

PHILIPPINES

Scale: 0 150 300 kilometres / 0 150 300 miles

---

a close-up encounter on a walk in the reserve. Mornings and evenings are best, and the monkeys tend to be more active in the mating season (February to May). Watch out for dive-bombing adolescents.

Père David's deer

One of the best reserves for seeing them is Dafeng Nature Reserve, a Ramsar Wetland of International Importance, southeast of Dafeng on the Yellow Sea coast in Jiangsu province – said to be the largest population in the world. Fly to Nanjing, and it's about 125km (78 miles) by road to Dafeng. Visit in June and July to see the rut.

red-crowned cranes

One of the top spots to see China's famous bird is the Yancheng Nature Reserve – another Ramsar Wetland of International Importance – beside the Yellow Sea. The reserve is a vast complex of coastal grassland, shrimp ponds, saltpans and commercial reedbeds. A third of the world's 2000 or so red-crowned cranes and 10 per cent of the world's black-faced spoonbills winter here. Binoculars are a must, as the birds are very wary of people. And prepare for cold and drizzle in winter. The extra rewards are a supporting cast of millions, including egrets, ducks, herons and curlews, and Père David's deer. There is a hotel ten minutes' drive from the reserve and 40km (25 miles) from the airport at Yancheng.

Captive red-crowned cranes at Zhalong.

To see the breeding display of the endangered red-crowned cranes, visit Zhalong Nature Reserve – also a Ramsar Wetland of International Importance – 30km (19 miles) from Qiqihaer in Heilongjiang province. The reserve is a breeding ground for red-crowned and white-naped cranes, and a stopover for Siberian and hooded cranes on migration.

The red-crowned cranes arrive in April and depart in September and October. Take powerful binoculars. Visit in spring before the reeds grow too high and when you have a chance of seeing their magical courtship dance. The reserve also has about 100 birds in captivity. Stay in Qiqihaer and take an hour's bus ride to the reserve.

whooper swans

Get close to whooper swans at Rongcheng in Shandong province. More than 2000 swans arrive here every November from Xinjiang Uygur autonomous region and Siberia. They overwinter until late March in four bays, collectively referred to as Roncheng Swan Lake. The locals feed them, and so you can get really close. The lake is about 28km (17 miles) from Weihai airport, and the closest town is Chengshan. The bays are packed with seaweed farms, and the village of Xukou (15 minutes' drive from the lake) has houses with traditional seaweed-thatched roofs.

LANDSCAPE WONDERS

Qiantang Jian tidal bore

If you want to see the biggest tidal bore in the world – a wall of water that can reach 9 metres (30ft) high thundering up the mouth of the Qiantang River from Hangzhou Bay – head to Yanguan, a small town about 38km (24 miles) northeast of Hangzhou. There is even a bore-watching festival, on the eighteenth day of the eighth lunar month, coinciding with the bore in September/October. Take a tour during the festival or get a bus to Yanguan from Hangzhou. Be prepared to run when the bore hits, and take waterproofs.

West Lake (Xi Hu), Hangzhou

This freshwater lake in central Hangzhou, Zhejiang province, offers a landscape resembling a traditional Chinese landscaped garden on a huge scale. It is split by two causeways, and the islands offer gardens, temples, parks and boat trips to the islands.

The Bund – Shanghai

The Bund (Zhongshan Dong Lu Road) is Shanghai's most famous street and one of China's most well-known cityscapes. Its name comes from an Anglo-Indian term for a muddy waterfront but is

Digging for cockles in one of the Swan Lake bays, the whooper swans watching on.

The Bund, Shanghai's famous financial street.

known locally as Wai Tan (outside beach). It is Shanghai's Wall Street, with the leading banks, trading houses, international businesses and hotels and a mix of 1930s Gothic, Renaissance and modern architecture that makes it feel like downtown New York by a river. During Shanghai's heyday, it was also a busy working harbour. Take a stroll down it at night when it is lit up. Or view it by boat from the Huangpu River, which will also give you views of Pudong Xinqu (the 'new area'), with its famous Oriental Pearl TV Tower.

Shuangtai Hekou Nature Reserve

The Liao River estuary in northeastern Liaoning province has the largest reedbed in China, protected as Shuangtai Hekou

Nature Reserve and another Ramsar Wetland of International Importance. It is on the eastern-Asia migration route for shorebirds and is a stopover for more than 400 critically endangered Siberian cranes on their southward migration. It is also the largest breeding site for Saunder's gull and the most southern breeding area for red-crowned cranes and grey seals.

Wuyi Shan Nature Reserve

This is a UNESCO World Heritage Site – cultural and natural – and lies in the far northeast of Fujian, close to the Jiangxi border. It boasts subtropical forests, cloudforests, waterfalls, craggy sandstone mountain peaks and winding river valleys, and is a wonderful place to relax for a few days. The scenic region consists of the Jiuqu (Nine Twists) River, which winds through

mountains and the 36 peaks. Two of the best walks are to Great King Peak and Heavenly Tour Peak, offering great views. Or raft leisurely down the Jiuqu River and view the stunning gorge scenery (trips run all year, and you can pick one at most places along the way).

Some 5000 species of animals have been recorded from the forests of Mt Wuyi, including endangered species such as Cabot's tragopan, Chinese black-backed pheasant and Chinese giant salamander. It was also one of the last haunts of the South China tiger. There are 44 plant species found only here. Reach it by bus, train or air from Fuzhou (or by air from Xiamen and other cities). A minibus will take you to the Wuyigong scenic area. High season is summer, when it is heaving with Taiwanese tourists.

Chinese tourists on Mt Wuyi taking in the view of the Jiuqu River gorge.

One of the many sandy beaches on tropical Hainan. Most are now given over to tourism.

Hainan Island

This is China's tropical tourist island. Head to Sanya for the sunshine beaches of Yalong Bay, Dadong-hai, Luhuitou peninsula and Tianya Haijiao. In the highlands of the southwest, wildlife hangs on in a few nature reserves. The lush mountain rainforests of Jianfengling Nature Reserve, 115km (71 miles) from Sanya, have a rich flora including orchids and birds such as the Hainan blue flycatcher. Bawangling Nature Reserve is home to the last few Hainan black-crested gibbons. In the centre, near Tongzha, is Five Fingers Mountain (Wuzhi Shan), which offers a challenging climb. The Sanya National Coral Reef

Nature Reserve in the southernmost part offers dives on what remains of the coral. Be warned, though: the island has been heavily exploited and is packed with tourists. Don't go between June and October – the wet season.

CULTURAL WONDERS

Fishing-lamp Festival

The fishermen of Penglai, about 65km (40 miles) northwest of Yangtai in Shandong province, celebrate this festival on the 16th of the first month of the lunar calendar (February or March) in the hope of a plentiful catch in the coming year and for safety on the sea. The fishermen and villagers

dress fish and pigs with red bows and bring them down to the boats. These offerings are thrown into the sea, and hundreds of firecrackers are set off. Penglai Pavilion, more than 1000 years old, offers great views of the boats and ocean.

Tai Shan

This peak in Shandong province is said to be the most sacred in China and is probably the most-climbed mountain on Earth. Being the most important of China's five Taoist mountains, it is almost a deity, attracting Taoist monks and a heaving mass of Chinese tourists. Along the trails are numerous pavilions and temples, but expect touts and hawkers rather than tranquillity. Tai Shan is 1545m (5069 ft) high and about 8km (5 miles) from the base to the summit, and with 6660 steps, it is a testing climb. Many think it's worth the effort: the saying goes that, if you climb Tai Shan, you'll live to be 100. Allow four or five hours to get up and about three to descend. You can either walk the whole way via the east or the west route or take a bus to where the paths meet at Zhongtiamen and walk or cable car the rest of the trail. It's possible to stay overnight at the Midway Gate to Heaven or at the top. The weather is very changeable, and so take warm and waterproof clothes and a torch. The best seasons are spring and autumn, but it's clearest from October. The town of Tian is the gateway to the mountain, and Jinan is the closest airport.

Mazu (goddess of the sea) Festival

For generations, local fishermen in China's southeastern coastal area have believed the sea goddess Mazu will ensure their safety and prosperity. On the anniversaries of her birth (day 23 of the third lunar month – around April) and her death (ninth day of the ninth lunar month – around October), celebrations occur at Mazu temples across southern China and Taiwan, and thousands of pilgrims, many from Taiwan, come to pay homage to her on Meizhou Island in the city

The testing climb up Tai Shan – 6660 steps.

of Putian, Fujian province, where the huge celebrations include Shaolin monk performers, dancing and processions.

Hakka roundhouses

These circular tulou (earth buildings) in Fujian province, built by the Hakka people in the Jin dynasty (AD 265–314) are fortresses big enough to house hundreds of people. Some are still inhabited, and others are maintained for tourism. Visit Zhencheng Lou, in a hamlet 5km (3 miles) north of Hukeng village – about an hour's bus ride from Yongding in southwest Fujian – which has a number of tulou houses that are still occupied.

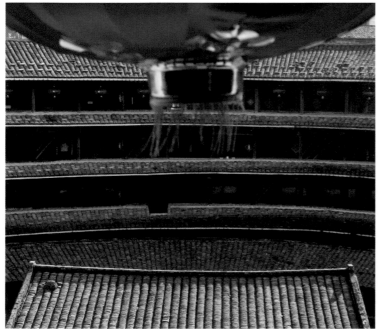

A Hakka roundhouse interior overlooking the temple roof.

further reading

Bonavia, J and Hayman, R, *Yangzi: The Yangtze River & the Three Gorges* (Odyssey Illustrated Guide, 2004).

Buckley Ebrey, P, *The Cambridge Illustrated History of China* (Cambridge University Press, 1996).

Chapman, G and Wang, Y, *The Plant Life of China: Diversity and distribution* (Springer-Verlag Berlin and Heidelberg GmbH & Co, 2002).

Grigsby, R, *China by Bike: Taiwan, Hong Kong, China's East Coast* (Mountaineers Books, 1994).

Harper, D, *Lonely Planet Country Guide: China* (Lonely Planet Publications, 2007).

Harris, R B, *Wildlife Conservation in China: Preserving the habitat of China's Wild West* (East Gate Books, 2007).

Laidler, L and K, *China's Threatened Wildlife* (Blandford, 1996).

Leader, P, Carey, G and Round, P, *A Field Guide to the Birds of China, Tibet and Taiwan* (Christopher Helm, 2008).

Lei Fumin and Lu Taichun, *China – Endemic Birds* (Science Press, 2006); available from NHBS Environment Bookstore.

Ma Jian, *Red Dust: A path through China* (Anchor Books, 2002).

MacKinnon, J and Hicks, N, *A Photographic Guide to the Birds of China Including Hong Kong* (New Holland Publishers, 2007).

MacKinnon, J, Showler, D, Phillipps, K, *A Field Guide to the Birds of China: Ornithology* (Oxford University Press, 2000).

Quinn, E, *Best of: Beijing* (Lonely Planet Publications, 2006).

Shapiro, J, *Mao's War Against Nature: Politics and the environment in revolutionary China* (Cambridge University Press, 2001).

Sheng Helin, Noriyuki Ohtaishi and Lu Houji, *Mammals of China* (China Forestry Publishing House, 1999); available from NHBS Environment Bookstore.

Wu Yipin, *Lifestyles of China's Ethnic Minorities* (Peace Book Company, Hong Kong, 1991).

CHAPTER 1

Catton, C, *Pandas* (Christopher Helm, 1990).

Clayre, A, *The Heart of the Dragon* (Collins/Harvill, 1984).

Lu Zhi and Schaller, G B, *Giant Pandas in the Wild: Saving an endangered species* (Aperture, 2002).

Lindburg, D and Baragona, K (editors) *Giant Pandas: Biology and conservation* (University of California Press, 2004).

Schaller, G B, *The Last Panda* (University of Chicago Press, 1994).

CHAPTER 2

Bonavia, J (revised by C Baumer), *The Silk Road: Xi'an to Kashgar* (Odyssey Publications, 2004).

Coggins, C, *The Tiger and the Pangolin: Nature, culture and conservation in China* (University of Hawaii Press, 2002).

Hare, J, *Lost Camels of Tartary: A quest into forbidden China* (Abacus, 1999).

Thubron, C, *Shadow of the Silk Road* (HarperCollins, 2007).

Xuncheng Xia, *Wondrous Taklimakan: Integrated scientific investigation of the Taklimakan Desert* (Science Press, Beijing, 1993).

CHAPTER 3

Buckley, M, *Bradt Travel Guide: Tibet* (Bradt, 2007).

Liu Wulin, *An Instant Guide to Rare Wildlife of Tibet* (China Forestry Publishing House, 1994); available from NHBS Environment Bookstore.

Mayhew, B and Kohn, M, *Lonely Planet Country Guide: Tibet* (Lonely Planet Publications, 2005).

Schaller, G B, *Tibet's hidden wilderness: Wildlife and Nomads of the Chang Tang Reserve* (Abrams, 1997).

CHAPTER 4

Bonavia, J and Hayman, R, *Yangzi: The Yangtze River & the Three Gorges* (Odyssey Illustrated Guide, 2004).

Booz, P R, *Yunnan: Southwest China's little-known land of eternal spring* (Verulam Publishing, 1987).

Chen Li and Zhang Jiangling, *Xishuangbanna: A nature reserve of China* (University of British Columbia Press, 1992).

Elvin, M, *Retreat of the Elephants: An environmental history of China* (Yale University Press, 2006).

Goodman, J, *Joseph F Rock and His Shangri-La* (Caravan Press, 2006).

Guan Kaiyun and Zhou Zhekun (editors), *Highland Flowers of Yunnan* (Science Press, 1998); available from NHBS Environment Bookstore.

Stotz, D F et al., *China: Yunnan, southern Gaoligongshan* (Chicago Field Museum of Natural History, 2003).

Wildlife of Yunnan in China (Chinese Academy of Sciences and Kunming Institute of Zoology, China Forestry Publishing House, 1999); available from NHBS Environment Bookstore.

Xie Jiru, *Bamboo Resources and Development Research of Yunnan* (China Forestry Publishing House, 1995); available from NHBS Environment Bookstore.

Xu Youkai and Liu Hongmao (editors), *Tropical Wild Vegetables in Yunnan* (China Science Press, 2002); available from NHBS Environment Bookstore.

http://drjosephrock.blogspot.com – In the Footsteps of Joseph Rock.

CHAPTER 5

Chuxing, H and Hong, L et al., *South China Karst 1* (Pensoft, 1998).

Viney, C, Phillips, K and Lam Chiu Ying, *The Birds of Hong Kong and South China* (Hong Kong Government Information Service, 2005).

Woodward, T, *Birding South-East China*; available from NHBS Environment Bookstore.

CHAPTER 6

Robson, C, *Birds of South-East Asia* (New Holland Publishers Ltd, 2005).

Sadovy, Y and Cornish, A, *Reef Fishes of Hong Kong* (University of Washington Press, 2000).

Wood, R E and Michael, A W, *Reef Fishes, Corals and Invertebrates of the South China Sea* (New Holland Publishers, 2002)

Woodward, T, *Birding South-East China*; available from NHBS Environment Bookstore.

Viney, C, Phillips, K and Lam Chiu Ying, *The Birds of Hong Kong and South China* (Hong Kong Government Information Service, 2005).

acknowledgements

This book and the television series it accompanies are inextricably linked. In researching the series, we collected much of the information which appears in the book, and in writing the book, we gained new insights which informed the content of the television programmes. Our thanks must therefore go to all those who contributed to the television series *Wild China* (shown in China as *Beautiful China*).

First we acknowledge with gratitude the efforts of the production team in Bristol and Beijing, who made this endeavour possible: BBC Natural History Unit head Neil Nightingale and executive producer Brian Leith, who steered the series towards a greater engagement with human and cultural stories; the Bristol team of Pauline Gates, Liz Toogood, Di Williams, Bridget Jeffery, Louise Davies, Becca Coombs, Alison Pilling and Claire Evans; and Poppy Toland and Shi Lihong, who provided research and translation support in China. We also thank our partners and families, who had to endure our frequent absences from home during the production.

Many Chinese ethnic minority communities, including the Hezhe, Ewenki, Mongolians, Kazakhs, Miao, Hani, Tibetans, Hui'an, Hakka and Dai, accepted our filming teams into their world and made us feel welcome, as did many Han Chinese communities. We thank them all for their tolerance, patience and hospitality.

We also thank the numerous scientists and conservation workers in China and elsewhere who shared their knowledge of Chinese natural history with us, in particular: Prof. Zhang Shuyi, Prof. Ablimit Abdukadir and Prof. Pan Wenshi of the Chinese Academy of Sciences; Prof. Zhu De-Hao of the Chinese Academy of Geological Sciences; Prof. Hu Defu of the Beijing Forestry University; Prof. Lu Zhi and Liu Yanlin of Beijing University; Prof. Baoguo Li of Northwestern University; Prof. Liang Congjie of Friends of Nature China; Sun Shan of Conservation International China; Prof. Xiong Kangning and Ren Xiaodong of Guizhou Normal University; Prof. Wu Xiaobing of Anhui Normal University; Qi Yun of Yunnan University; Jason Lees and the crew at Haiwei Trails; Dr Andreas Wilkes of the Mountain Institute; Josef and Minguo Margraf of the Tianzi Centre; Dr Yin Shaotin; Dr Li Bo of the Centre for Biodiversity and Indigenous Knowledge at Qi Yun; Long Yong Chen of the Nature Conservancy; Dr Pete Winn; John Corder; Dr Craig Fitzpatrick of TRAFFIC East Asia; Prof. Ding Yuhua at Dafeng Deer Reserve; Liu Wulin of Tibet Forestry Institute; Ciren Yangzong of Tibet University; Dega of Tibet TV; Rich Harris of the University of Montana; Baozhong Lu of Shaanxi Crested Ibis Station; Dr Rao Dingqi of the Kunming Institute of Zoology; Nina Jablonski of the Department of Anthropology, California Academy of Sciences; Dr Bill Bleisch of Fauna and Flora International; Dr Roger Luo of IFAW Asian Elephant Conservation; Dr Philip McGowan Director of the World Pheasant Association; Mr Jin and staff at Dontang Nature Reserve; Mr Bena Smith and Dr Lew Young of Mai Po Nature Reserve; the staff at the Yanchen, Zhalong, Bawangling, Caohai, Mayanghe and Changqing Nature Reserves; Alison Foot; Libiao Zhang; Fan Peng Fei; Cyril Grueter; Prof. Zhang Zhengwang; Mrs Yi Ran; Mr and Mrs Qu; Dr Yvonne Sadovy; and Dr Lindsay Porter.

We thank Shirley Patton for having the confidence in the series to commission the book and Bobby Birchall for his thoughtful design. Finally, we thank our editor Rosamund Kidman Cox, whose patience and skill have turned our motley contributions into a proper book.

picture credits

index

index

This book is published to accompany the
television series entitled Wild China, first
broadcast on BBC2 in 2008.

Published in the United States in 2008 by Yale
University Press
Published in the United Kingdom in 2008 by
BBC Books, an imprint of Ebury Publishing.
A Random House Group Company

Commissioning Editor Shirley Patton
Project Editor Rosamund Kidman Cox
Designer Bobby Birchall, Bobby&Co Design
Production David Brimble
Cartography Encompass Graphics

Colour origination by GRB Editrice, London
Printed and bound in the UK by Butler and
Tanner

Library of Congress Control Number:
2008921278
ISBN 978-0-300-14165-8 (pbk.: alk. paper)

A catalogue record for this book is available
from the British Library.

This title has been printed on
Greenpeace-approved Forest Stewardship
Council–certified paper.

10 9 8 7 6 5 4 3 2 1